Math Activities with

DOMINOES

Helene Silverman
Sandy Oringel

ETA/Cuisenaire®
www.etacuisenaire.com
800-445-5985

Managing Editor: Alan MacDonell
Acquisitions Manager: Doris Hirschhorn
Development Editor: Harriet Slonim
Design Director: Phyllis Aycock
Cover Design and Illustration: Tracey Munz
Text design, line art, and production: Fiona Santoianni

© Copyright
ETA/Cuisenaire®
500 Greenview Court, Vernon Hills, IL 60061-1862

2 3 4 5 6 7 8 9   06 05 04 03 02 01

Table of Contents

Introduction

Math Activities with Dominoes provides hands-on, student-centered math tasks for children in grades three through eight. The activities encourage students to investigate basic math concepts, construct their own mathematical meanings, and communicate their ideas. As they explore with dominoes, students have the opportunity to become confident problem solvers, developing mathematical thinking abilities and assimilating concepts. This is consistent with the vision proposed by the National Council of Teachers of Mathematics (NCTM) in the *Curriculum and Evaluation Standards for School Mathematics*. The first four standards involve problem solving, communication, reasoning, and making mathematical connections.

About Dominoes

Dominoes are available in double-six and double-nine sets. Each domino piece is divided into two squares, called *faces*. Each face has a value determined by a number of spots, called *pips*.

The double-six set of dominoes consists of 28 pieces. Each piece has from 0 to 6 pips on each of its faces. Twenty-one dominoes, called *singles,* have two different numbers of pips on each face. A domino face with no pips is called a *blank,* or *zero.* Seven of the pieces, called *doubles,* have equal numbers of pips on their faces.

Singles   **Doubles**

The double-six piece is the domino with the greatest value. The double-zero, or double-blank, piece is the domino with the least value.

Double-six   **Double-zero/Double-blank**

The values 0 to 6 are represented in each domino *suit*. There are seven suits, each with seven dominoes. Except for the doubles, each domino is a member of two different suits. For example, the zero suit contains the following dominoes:

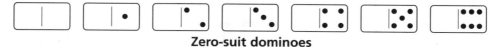

Zero-suit dominoes

The three-suit contains these dominoes:

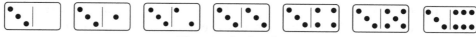

Three-suit dominoes

Notice that the 0–3 domino in the zero suit is the same as the 3–0 domino in the three suit, and so this domino is in both suits.

The complete 28-piece set of double-six dominoes have pips that represent the suits from 0 to 6.

The complete 55-piece set of double-nine dominoes have pips that represent the suits from 0 to 9. The seven-, eight-, and nine-suit dominoes added to the double-six set make up the double-nine set.

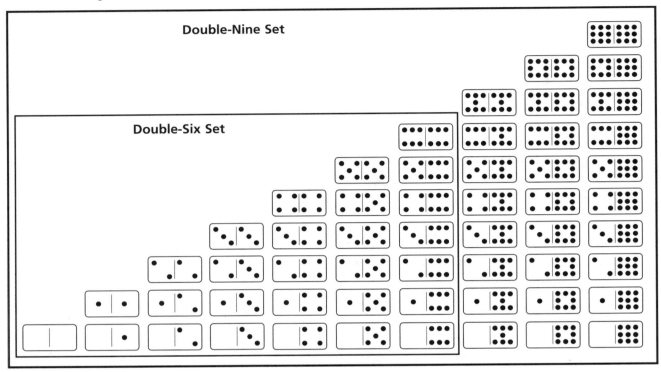

About the Activities

In preparation for working with the activities, multiple sets of dominoes should be acquired so that students can work in pairs or small groups. Sets of double-six dominoes in a variety of colors would be useful for some activities. Most activities can be modeled for the class using large domino cards or an overhead projector. They can be modeled for a small group at a desk or table. After you explain an activity, students can work on it in pairs or small groups, either at their desks or at a math center. Students may wish to work on some activities at home.

In order to keep records of their trials, solutions, and strategies, students should have pencils and paper available. Alternatively, you may want them to record their work in their math journals.

Mats, accompanying many of the activities, serve as work areas for some activities and game boards for others. Students will want to play the games over and over again. You may want to provide for this by duplicating the game boards and then coloring and laminating them for repeated use. Consider making additional copies of all the mats so that students can take them home and share the activities with their families.

To help students keep track of the dominoes as they use them in certain activities, you may want to provide them with the following tracking record. (This tracking record can be adjusted for use with double-nine dominoes.)

Tracking Record (Double-Six Dominoes)

0	0	0	0	0	0	0	1	1	1	1	1	1	2	2	2	2	2	3	3	3	3	4	4	4	5	5	6
0	1	2	3	4	5	6	1	2	3	4	5	6	2	3	4	5	6	3	4	5	6	4	5	6	5	6	6

The activities are organized in five chapters, each emphasizing a major mathematical topic: number sense, logic, patterns, number theory, and computation. While each of the activities develops a specific mathematical concept, the emphasis is on problem solving, observation, classification, and estimation.

Domino codes appear at the top of each page of teachers' notes. They indicate whether the activity is intended to be implemented with double-six or double-nine dominoes or both.

The teachers' notes are geared to help both teachers and students get the most from the activities. Each page of notes is organized as follows:

Task gives the main objective, skills developed, or purpose.

Setup indicates the materials needed and suggested groupings of children.

Start-up describes the activity, procedure, or game rules.

Discussion focuses on how children generally respond to the task and ways to address questions that may arise. Sample solutions are often provided.

Keep up provides ideas for keeping the task on target.

Wrap-up suggests critical-thinking questions for class discussion or for use in making journal entries.

Follow-up expands the task for the more experienced students.

About the Activity Features

- Each activity encourages students to solve problems through the use of process skills such as *describing, sorting, matching, ordering, listing, counting, comparing, patterning,* and *generalizing.* Some activities can be solved in different ways; some have different solutions. Most of the games can be played using various strategies for winning. Upon completing an activity students may be asked to describe how they solved a problem or what strategy they used in a game. Such questioning helps students build confidence in their reasoning.
- Investigation is the emphasis of all the activities. Students are encouraged to record their work and share their ideas. Extensions and variations of the activities are developed in many of the Keep-up and Follow-up features and through the probing Wrap-up questions.
- Activities are designed to be open ended and can easily be adapted to meet the needs of any student group. Game rules can be altered to simplify an activity or to make it more challenging. Students are encouraged to offer suggestions for how to change the rules of a game.
- The activity mats and game boards are designed to be reproduced.
- Blackline masters for domino sets can be found on pages 78–80.
- Calculators may be used with those activities that involve many computations. Calculators free students to focus their attention on meeting the challenges of the activities.

Getting Started with Dominoes

Task   Students are introduced to the characteristics and terminology that describe a set of dominoes. Then they sort and graph the set.

Setup   A pair or small group of students uses one set of double-six dominoes. Number cards (0 to 12) found on page 59 are optional.

Start-up   Display a domino and ask students to share what they know about dominoes. As students point out features, encourage a variety of responses and the use of the domino terminology. Show two dominoes and ask students to tell how the faces of these dominoes are alike and/or how they are different. Do this several times with different pairs of dominoes.

Students sort a set of dominoes in any manner they choose and then describe how they sorted and what they notice about the set. They arrange their dominoes in rows or columns to form a graph. They may use number cards to label their rows or columns.

Discussion   The domino terminology explained in the first part of the Introduction, "About Dominoes," should become an integral part of students' vocabulary as they work with dominoes. As students sort the dominoes, they will notice likenesses and differences between the individual pieces. As they graph the dominoes, they should notice that each domino in a set is different from all the others. Students may graph a double-six set according to suit in columns numbered from 0 to 6. They will notice that most dominoes can be placed in two different columns as indicated by their two faces. Alternatively, students may graph the dominoes according to value in columns numbered from 0 to 12. They will notice that the number of dominoes of different values varies. There is one domino for each of the values 0, 1, 11, and 12 and as many as four dominoes with the value 6. Familiarity with the set of dominoes will be helpful in planning strategies for games and other activities.

Keep up   Have students sort and graph their set of dominoes in a way that is different from the way they chose at first. Then have them choose one of the graphs to record and describe.

Wrap-up   Key questions for discussion or response in journals:
- How did you graph your dominoes?
- What are the likenesses and/or differences between the columns or rows in your graph?

Follow-up   Have students sort and graph a set of double-nine dominoes. Have them record and describe their graph.

Number Sense

Students estimate, classify, and sort dominoes to help them investigate different domains of number. After reading dominoes as two-digit numbers, proper fractions, and decimals, students gain practice in working with inequalities and in identifying equivalent values.

Task   Students play a game in which they pick dominoes, identify them as two-digit numbers, and then find the sum of the numbers.

Setup   A pair or small group of students uses one set of double-six dominoes.

Start-up   Hold up a single domino horizontally and point out that it can be read as two different two-digit numbers. The 3–4 domino, for example, can be read either as 34, with the greater number of pips in the ones place, or as 43, with the greater number of pips in the tens place.

3 4   4 3

The dominoes are placed face down on the playing area. Players take turns picking a domino, deciding on its value as a two-digit number, and recording the number. Each player keeps a running sum of the numbers named for five picks, trying to reach a sum of 100 without going over it. One domino may be rejected if a player thinks either of its values is so high that it would bring the sum over 100 or so low that it would not bring the sum close enough to 100. Each player may reject only one domino. Even a rejected domino is considered to be one of the five picks. The player with the sum closest to 100 without going over it wins.

Discussion   Students will notice that choices they made early in the game can affect the sum. With each pick, they decide whether to use the domino or reject it. They either calculate or estimate the impact of this decision on the next sum. Many students get as close as possible to 100 with their first four dominoes, knowing they may reject their fifth domino.

Sample round for two players:

PLAYER 1				PLAYER 2			
PICK	DOMINO		SUM	PICK	DOMINO		SUM
1st	3–4	Keep	34	1st	4–2	Keep	42
2nd	5–5	Reject	—	2nd	1–5	Keep	57
3rd	3–1	Keep	65	3rd	1–4	Keep	71
4th	2–6	Keep	91	4th	0–6	Keep	77
5th	0–4	Keep	95	5th	2–4	Reject	—

Player 1, with a sum of 95, wins.

Keep up   Have students look again at the dominoes they picked in a game and determine either the highest or the lowest score possible with those picks. Remind them that they may reject any one of their five picks in determining this score.

Wrap-up   Key questions for discussion or response in journals:

- Each time you picked a domino, how did you decide whether to read it with the greater number of pips in the tens place or in the ones place?

- When did you decide to reject a domino? Explain.

Follow-up   Have students play another game, this time playing so that the winner is the one whose sum is closest to 100 even if it goes beyond 100. In another version of the game, instead of having players add to get a sum closest to 100, have them start with 100 and subtract to get a sum closest to 0.

Close to 2

Task   Students play a game in which they pick dominoes, identify them as fractions, and find the sum of the fractions.

Setup   A small group of students uses one set of double-six dominoes with the zero-suit dominoes and the 1–5, 2–5, 3–5, and 4–5 dominoes removed.

Start-up   Hold up a single domino vertically and point out that it can be read as a fraction, with the face having fewer pips (the numerator) above the face having more pips (the denominator). For example:

$$\frac{1}{3} \qquad \frac{3}{4}$$

Players each pick two dominoes from the facedown set. They turn one face up and stand up the other so that only they can see it. Players mentally add the values of their dominoes, trying for a sum of 2 without going over it. A player with a low sum may pick one or more dominoes in hopes of bringing the sum closer to 2. Once a domino is picked it is turned face up and must be included in the hand. When players are satisfied with their hands or when one of their sums exceeds 2, they compare hands. The player whose sum is closest to 2 without going over it gets 1 point. (In a *tie* each player gets 1 point.) The dominoes used for the round are set aside. Play continues until no more dominoes are left. After a predetermined number of rounds, the player with the greatest number of points wins.

Discussion Encourage students to round their fractions and estimate the sums. Students unable to do this mentally may need paper and pencil. Be sure students understand that, whenever they pick an extra domino, they risk bringing the sum beyond 2.

Two sample hands:

$$\underset{\text{Down} \quad \text{Up}}{\boxed{}\;\boxed{}} = 1\frac{1}{3} \longrightarrow \overset{\text{Pick}}{\boxed{}} = 1\frac{1}{3} + \frac{2}{3} = 2$$
$$\frac{4}{6} = \frac{2}{3}$$

$$\underset{\text{Down} \quad \text{Up}}{\boxed{}\;\boxed{}} = \frac{2}{3} + \frac{4}{6} = 1\frac{2}{6} \longrightarrow \overset{\text{Pick}}{\boxed{}} \longrightarrow 1\frac{2}{6} + \frac{3}{6} = 1\frac{5}{6}$$

Keep up   Have students record their fractions and explain how they calculated their sums.

Wrap-up   Key questions for discussion or response in journals:

- Which two dominoes could you pick at the beginning of a round that would make you an instant winner? Explain.

- When you decided to pick more dominoes for your hand were the dominoes you picked helpful or harmful? Explain.

Follow-up Increase the difficulty of the game by returning the 1–5, 2–5, 3–5, and 4–5 dominoes to the set. Have students compare this game to their previous games.

On or About

Task   Students identify dominoes as fractions. They compare the values of their domino fractions by sorting them on a number line.

Setup   A pair of students uses an "On or About" mat and one set of double-six dominoes with the zero suit removed. Fraction manipulatives are optional.

Start-up   Hold up a domino vertically as you point out that it may be read as a fraction, with the face having fewer pips (the numerator) above the face having more pips (the denominator).

The dominoes are placed face down on the playing area. Students take turns picking a domino, determining its value as a fraction, and placing it on the mat according to whether its value is *about 0, about ½,* or *about 1.* Students each pick and place three dominoes, evaluate how they placed them, make changes if necessary, and record the placement. Then each pair mixes up its six dominoes, trades them with another pair, and sorts the other pair's dominoes on the mat. Pairs compare results.

Discussion   Although students quickly agree about where to place dominoes with values equivalent to ½ and 1, they may disagree about where to place other dominoes. For example, one member of a pair may want to place the 1–4 domino in the *about 0* box. The other member of the pair may say it belongs in the *about ½ box.* In this case, either of the two positions would be legitimate! Such discussion builds students' understanding of the meaning of fractions. Many students would benefit from the use of fraction manipulatives—various shapes divided into fractional parts—to help them in placing their dominoes.

Keep up   After they have placed their dominoes, have students move them down onto the number line where they can order them from least to greatest according to the fractions they represent. Then have students record the fractions in order.

Wrap-up   Key questions for discussion or response in journals:

- How did you decide in which box to place a domino?
- Which dominoes were the most difficult to place? Explain.

Follow-up   The number line may be extended to include markings for 1½ and 2. Pairs can then repeat the activity. This time have them pick two dominoes at once, identify each as a fraction, estimate the sum of the fractions, and place the dominoes together on the extended number line according to their sum.

On or About

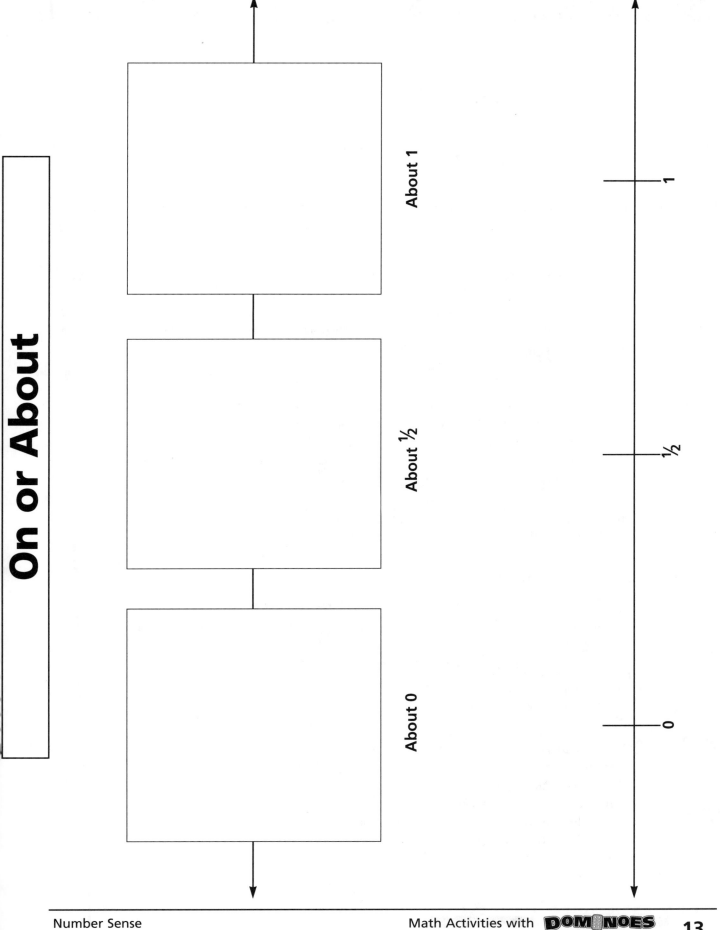

About 1

About ½

About 0

1

½

0

Follow the Leader

Task   Students read dominoes as fractions and compare their values to the value of a single domino fraction.

Setup   A small group of students uses one set of double-six dominoes with the zero suit removed and a "Follow the Leader" mat.

Start-up   Hold up a domino vertically as you point out that it may be read as a fraction, with the face having fewer pips (the numerator) above the face having more pips (the denominator).

The dominoes are placed face down on the playing area. One domino is turned over, identified as the "Leader," and placed at the top of the mat. Students take turns picking a domino, determining its value as a fraction, and deciding whether the value is *less than, equal to,* or *greater than* the value of the Leader. They place the domino in the corresponding column on the mat. Students continue picking and placing dominoes until no more are left.

Discussion   Students will notice that each domino has a specific location when compared to the Leader. They may need to express the fraction values of some of their dominoes in lowest terms before they can compare them to the Leader. After the Leader is placed, and before other dominoes are picked, you may want to have students predict which column is likely to have the fewest and/or the most dominoes.

Keep up   Provide additional copies of the mat on which students can record the positions of the dominoes they pick for various Leaders.

Wrap-up   Key questions for discussion or response in journals:

- How did you decide on the column in which to place a domino?

- Imagine placing all the dominoes for one Leader and then changing the Leader. Why might the positions of the dominoes have to change for the new Leader? Why might they *not* have to change?

- Which dominoes would always belong in the same columns no matter which Leader was chosen? Explain.

Follow-up   Have a pair of students, each with his or her own mat, share a set of dominoes with the zero suit removed. Each picks a Leader and places it on his or her mat. Then they take turns picking and placing dominoes until no more are left to pick. Students compare the dominoes on their own mats against their partner's Leader. They take turns selecting one domino from their mat and placing it on their partner's mat according to the Leader on that mat. Then students discuss how and why their dominoes change columns as they are moved from one mat to another.

Follow the Leader

LEADER

LESS THAN	EQUAL TO	GREATER THAN

Task   Students play a game in which they first identify dominoes as fractions and then sort the dominoes according to whether or not they represent lowest-terms fractions.

Setup   A small group of students uses one set of double-nine dominoes with the zero suit removed. Each student uses a "Box Score" mat.

Start-up   Show students how to read dominoes as fractions by holding up a domino vertically with the face having fewer pips (the numerator) above the face having more pips (the denominator).

The dominoes are placed face down on the playing area. Players take turns picking a domino and deciding whether or not the fraction it represents is in lowest terms. If it is in lowest terms, the player places it on the mat in the box that matches the value of the numerator. If it is not in lowest terms, the player states the lowest-terms fraction and places the domino in the 10-box. After a player places a domino, the rest of the group evaluates whether or not it has been placed correctly. If it has been incorrectly placed, the domino is remixed in the pile and the player loses that turn. Play continues until players have picked all the dominoes. Each player determines his or her score by multiplying the number of dominoes in each box by the box number and then adding to find the sum of the products. After a predetermined number of rounds, the player with the highest score is the winner.

A sample round for one player:

Discussion   Some students may recognize that a fraction is not in lowest terms but may be unable to simplify it mentally. Provide these students with paper and pencil for calculating as well as for determining their scores.

Wrap-up   Key questions for discussion or response in journals:

- Each time you picked a domino, how did you decide where to place it on the mat?

- In which box did you place the most dominoes? Why?

Follow-up   Students familiar with improper fractions—fractions with values equal to or greater than 1—can play the following version of the game. Be sure students know how to read a domino as an improper fraction by holding it vertically with the face having more pips (the numerator) above the face having fewer pips (the denominator). Students pick dominoes, placing those that represent whole numbers into boxes 1 – 8 and those that represent mixed numbers into the 10-box. Have students discuss the differences between this game and the original game.

BOX SCORE
1 x 3 = 3
2 x 2 = 4
3 x 1 = 3
4 x 1 = 4
5 x 1 = 5
6 x 0 = 0
7 x 0 = 0
8 x 1 = 8
10 x 3 = 30
TOTAL: 57

Box Score

1	2	3	4	10
5	6	7	8	

Task   Students play a game in which they capture fractional parts of regions. The goal is to be the one who captures the greater number of whole regions.

Setup   Each pair of students uses one set of double-six dominoes (with the zero suit removed) in a paper bag, a "Domination" mat, and two markers or crayons of different colors.

Start-up   Hold up a domino vertically so that the face having fewer pips is above the face having more pips. Point out that this domino may be read as a fraction whose value is less than 1. Then hold up several doubles and point out that each of these dominoes may be read as a fraction whose value is equal to 1.

One player picks a domino from the bag and states its value as a fraction whose value is either less than or equal to 1. The player captures the corresponding number of parts of one of the regions on the mat by coloring them and then returns the domino to the bag. Players take turns picking, identifying the values of the domino fractions, and using their own color to capture parts of regions with a goal of filling entire regions with their own color. On each turn, a player must color as many parts of one region as possible. If that region becomes filled, the player may complete the turn by coloring parts of another region. The player who captures the most regions is the winner.

Discussion   As the game progresses, students increasingly use what they know about equivalent fractions to capture regions. For example, a student who has already captured two parts of a region divided into sixths might pick the 2–3 domino ($\frac{2}{3}$). Knowing that $\frac{2}{3}$ is equivalent to $\frac{4}{6}$, instead of coloring two parts of a thirds region, the student may choose to color the remaining four parts of the sixths region and thereby capture the entire region. Some students may use a blocking strategy, capturing parts of regions that their opponents have already begun.

Keep up   Students may agree on a change in the game rules. One such change might be to allow players to bank some fractional parts from one pick for use on a subsequent turn. For example, a student who picks the 3–4 domino ($\frac{3}{4}$) and wishes to color only two parts of a fourths region ($\frac{2}{4}$) in order to complete and capture that region may do so, recording the remaining $\frac{1}{4}$ and banking it so it can be combined with another fraction on a later pick.

Wrap-up   Key questions for discussion or response in journals:

- Which regions were easiest to capture? Which were most difficult? Explain.
- What strategy did you use to play? Describe how successful it was.

Domination

1/1	1/1

1/2 1/2	1/2 1/2	1/2 1/2	1/2 1/2

1/3 1/3 1/3	1/3 1/3 1/3	1/3 1/3 1/3	1/3 1/3 1/3

1/4 1/4 1/4 1/4	1/4 1/4 1/4 1/4	1/4 1/4 1/4 1/4	1/4 1/4 1/4 1/4

1/5 1/5 1/5 1/5 1/5	1/5 1/5 1/5 1/5 1/5	1/5 1/5 1/5 1/5 1/5	1/5 1/5 1/5 1/5 1/5

1/6 ×8	1/6 1/6 1/6 1/6 1/6 1/6	1/6 ×8	1/6 1/6 1/6 1/6 1/6 1/6

Math Activities with **DOMINOES**
© ETA/Cuisenaire®

Strive for 5!

Task   Students play a game in which they pick dominoes, identify them as whole numbers and/or decimal numbers, and find the sum of the numbers.

Setup   A small group of students uses one set of double-six dominoes.

Start-up   Hold up a single domino horizontally, pointing out that in each of two positions, it can be read as two different decimal numbers. The 2–3 domino, for example, can be read either as 2.3 or 0.23 or as 3.2 or 0.32.

2.3 or 0.23   **3.2 or 0.32**

The dominoes are placed face down on the playing area. Players take turns picking a domino, deciding how to read it as a decimal, and recording the decimal. Each player keeps a running sum of the decimals named for five picks, trying to reach a sum of 5 without going over it. One domino may be rejected if a player thinks each of its four possible values is so high that it would bring the sum over 5 or so low that it would fail to bring the sum close enough to 5. Each player may reject only one domino. Even a rejected domino is considered as one of the five picks. The player with the sum closest to 5 without going over it wins the game.

Discussion   Students will notice that the choices they made early in the game can affect the sum. With each pick, they decide whether to use the domino or reject it. They calculate or estimate the impact of this decision on the next sum. Many students get as close as possible to 5 with their first four dominoes, knowing they may reject their fifth domino.

Sample round for two players:

	PLAYER 1				PLAYER 2		
PICK	DOMINO		SUM	PICK	DOMINO		SUM
1st	0.34	Keep	0.34	1st	2.4	Keep	2.4
2nd	0.55	Keep	0.89	2nd	1.5	Keep	3.9
3rd	1.3	Keep	2.19	3rd	0.45	Keep	4.35
4th	2.6	Keep	4.79	4th	0.66	Reject	—
5th	0.04	Keep	**4.83**	5th	0.60	Keep	**4.95**

Player 2, with a sum of 4.95, wins.

Keep up   Students look again at the dominoes they picked in a game and determine either the highest or the lowest score possible with those picks. Remind them that they may reject any one of their five picks in determining this score.

Wrap-up   Key questions for discussion or response in journals:

- Each time you picked a domino, how did you decide whether to read it as a whole number and a number of tenths or as a number of hundredths?

- When did you decide to reject a domino? Explain.

Follow-up   Have students play another game, this time playing so that the winner is the one whose sum is closest to 5 even if it goes beyond 5. In another version of the game, instead of adding to get a sum closest to 5, players start with 100 and subtract to get a sum closest to 0.

Logic

Students classify, order, and compare the numerical values of dominoes. They group dominoes according to their attributes and then arrange the groups in a variety of geometric formats and Venn diagrams.

Blackout

Task Students play a game in which they try to cover a domino-patterned game board with matching dominoes. The goal is to be the first to completely cover the board.

Setup Each pair of students uses one set of double-six dominoes and a "Blackout" mat. (A set of double-nine dominoes may be used for the Follow-up game.)

Start-up The mat is cut in half so that each student has a game board. The dominoes are turned face up on the playing area. Players take turns looking for a domino whose two faces exactly match adjacent squares on their game board. When they find that domino they use it to cover the matching faces on the board. On any turn, instead of picking, a player may choose to remove a domino from the game board or to move a domino to another position. The winner is the first player to cover his or her game board completely. (If neither game board is completely covered but students are unable to make a play the game ends, and the player with the fewest uncovered faces wins.)

Discussion As students match dominoes to their game boards, they must be sure to place them so that the pip arrangements are oriented in exactly the same ways. They should take particular care to match the ways in which two, three, and six pips appear on the dominoes with the ways they appear on the game board.

Some students may notice how the symmetry of the pip patterns affects playing a domino. When a student is declared the winner, the game board should be checked to be sure that the faces of the dominoes are correctly oriented.

Not This   But This

One possible solution:

Gameboard A    **Gameboard B**

It is possible to cover both game boards with one set of dominoes simultaneously.

Keep up Either as a warm-up to familiarize students with the course of play, or as a way to help students develop playing strategies, you may want to have them play *Blackout Solitaire*. In this game, each student works alone with a complete domino set from which he or she seeks out the dominoes needed to completely cover both game boards.

Wrap-up Key questions for discussion or response in journals:

- How did you know when two faces that are next to each other on the game board could not possibly be on the same domino?
- Where on the game board is it best to begin placing dominoes? Explain.

Follow-up Additional Blackout-mat game boards may be generated for double-six and/or double-nine domino sets. Just put facedown dominoes on a copy machine and manipulate them into different-sized rectangular configurations. Cover the configurations with a sheet of paper and make the copy. Use a felt marker to darken any white gaps that appear on it. Then copy the copy to produce the finished game boards.

Blackout

Gameboard A

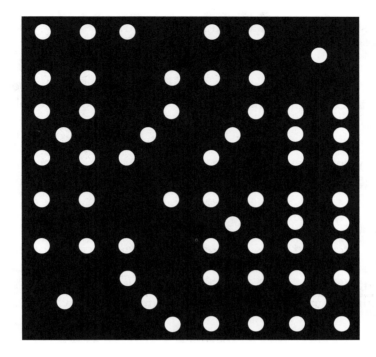

Blackout

Gameboard B

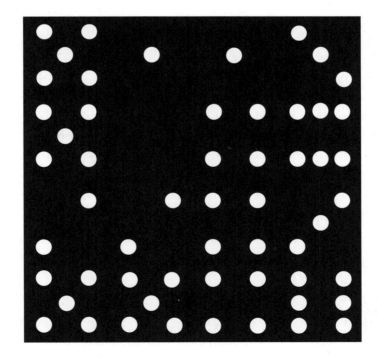

Double-Circle Sets

Task   Students sort dominoes according to their attributes on a Venn diagram formed by two overlapping circles.

Setup   A small group of students uses one set of double-six dominoes, a "Double-Circle Sets" mat, and a set of "Circle Sets Attribute Cards." (A set of double-nine dominoes is needed for the Follow-up.)

Start-up   You may want to review the meanings of the terms on the Attribute Cards by having students identify at least one domino associated with each card.

The group randomly chooses two cards and puts them on the mat. Each student picks four dominoes from the facedown set. Students take turns placing one of their dominoes according to its attribute—in one circle or the other or in the intersection of the circles. For each domino placed, the student picks another from the pile. Any domino that does not belong within the circles must be put down outside the circles. Play continues until no more dominoes are left to play.

Discussion   With two Attribute Cards on the mat, you may want to model the placement of a few dominoes, especially those that belong in the intersection of the circles. Point out that depending on the cards picked, there may be times when no dominoes belong in the intersection. With experience, as soon as the Attribute Cards are chosen, some students will be able to tell whether or not the intersection will remain empty.

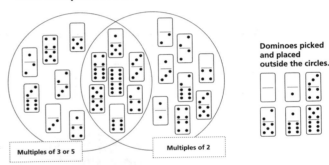

One sample solution:

Multiples of 3 or 5   Multiples of 2

Dominoes picked and placed outside the circles.

Keep up   Students may prepare additional cards for other attributes. They may also repeat the activity as a competitive game by receiving 3 points for each domino placed inside a circle, 5 points for each domino placed in the intersection of the circles, and 1 point for each placed outside the circles.

Wrap-up   Key questions for discussion or response in journals:
- Each time you picked a domino, how did you decide where to place it?
- Which pairs of Attribute Cards would cause the intersection of the circles to remain empty? Explain.

Follow-up   Have students repeat the activity using a set of double-nine dominoes. You may want to have them write additional attributes on the blank cards.

Double-Circle Sets

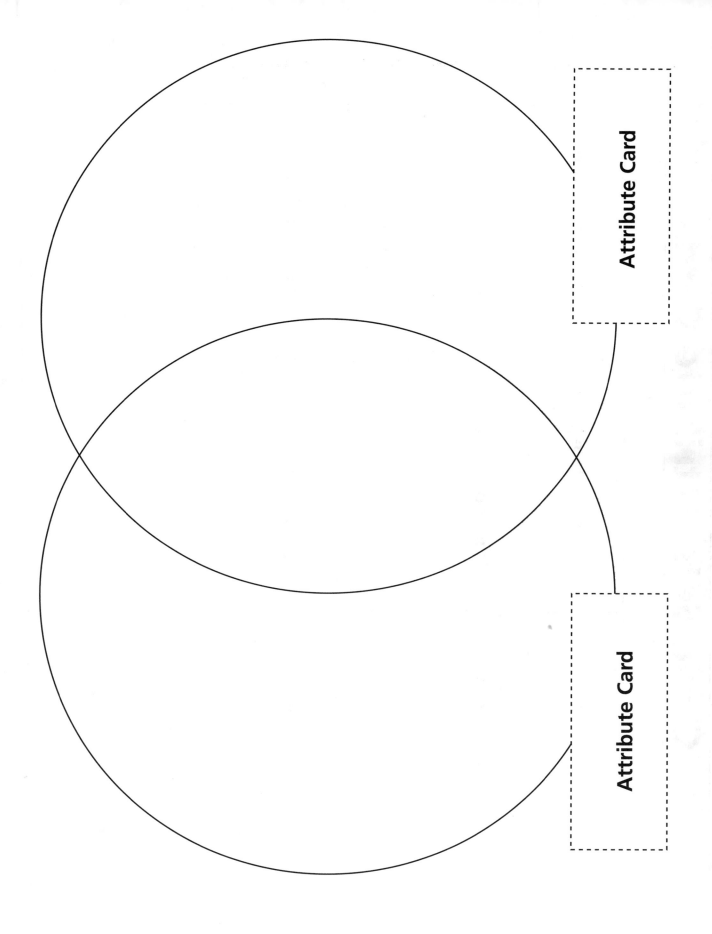

Attribute Card

Attribute Card

Circle Sets Attribute Cards

Odd-number sums	At least one face blank	Doubles
Sums equal to 7	Sums equal to 6	Even-number sums
Multiples of 2	Factors of 36	Factors of 24
Prime numbers	Multiples of 3 or 5	Multiples of 3
Square Numbers	Odd-number sum *not a factor of 24*	Even-number sum *not a multiple of 4*
		Triangular Numbers

Triple-Circle Sets

Task   Students sort dominoes according to their attributes on a Venn diagram formed by three overlapping circles.

Setup   A small group of students uses one set of double-six dominoes; three pieces of string, each about 4 ft long and tied into a loop; and a set of "Circle Sets Attribute Cards." (A set of double-nine dominoes is needed for the Follow-up.)

Start-up   Students place the string loops on the playing area so that they form three overlapping circles. They randomly choose three Attribute Cards, putting one on the edge of each circle. Each student picks five dominoes from the facedown set. You may want to help students place a domino according to its attribute—in one circle or another or in an intersection of two or three circles. For each domino placed, the student picks another from the pile. Any domino that does not belong within the circles must be put down outside the circles. Play continues until no more dominoes are left to play.

Discussion Those students who have not had experience with Venn diagrams may continue to need help in placing dominoes—especially those dominoes with more than one attribute. Be sure students understand why a specific domino with multiple attributes belongs in an intersection. Point out that depending on the attributes picked, it may be that no dominoes belong in an intersection. As soon as the cards are put in position, more experienced students will be able to identify an intersection that will remain empty.

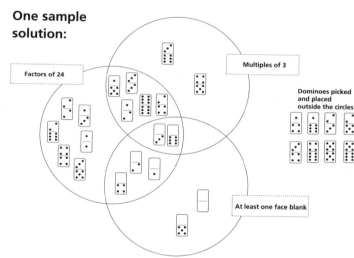

One sample solution:

Factors of 24

Multiples of 3

Dominoes picked and placed outside the circles.

At least one face blank

Keep up   To play competitively, students may receive points for their dominoes—1 point for each placed inside a circle, 2 points for each placed in an intersection of two circles, and 3 points for each placed in the intersection of all three circles. Students receive no points for dominoes placed outside the circles.

Wrap-up   Key questions for discussion or response in journals:

 • Each time you picked a domino, how did you decide where to place it?

 • Which groups of three Attribute Cards used together would cause the intersection of all three circles to remain empty? Explain.

Follow-up Have students repeat the activity using a set of double-nine dominoes. Encourage them to write additional attributes on the blank cards.

Next Hex

Task   Students arrange fourteen dominoes on a hexagonal grid so that no two dominoes with consecutive values are in adjacent cells.

Setup   Each pair of students uses a "Next Hex" mat and these dominoes taken from a double-nine set: 0–1, 0–2, 0–3, 0–4, 0–5, 0–6, 3–4, 2–6, 3–6, 5–5, 5–6, 6–6, 6–7, 7–7. (A double-six set may be used if the dominoes with values of from 1 to 12 are removed and two of the end cells are crossed off the mat.)

Start-up   Students position the dominoes on the mat, one to a cell, so that no two dominoes with consecutive values are in cells that touch. For example, neither the 3–6 domino nor the 5–6 domino, with values of 9 and 11 respectively, may be placed in cells that touch the cell with the 5–5 domino because they would then be touching a cell with a consecutive value, 10.

Discussion   As students examine the grid, they should notice that some hexagonal cells touch three other cells, some touch four, and some touch as many as six. Students should consider the possible advantages of beginning by placing dominoes at various starting points, such as in those cells that touch six other cells or at either end of the grid. For students having difficulty placing dominoes, it may help to have them arrange the fourteen dominoes in order of value, from the one with the least value (0–1, with a value of 1) to the one with the greatest (7–7, with a value of 14).

One sample solution:

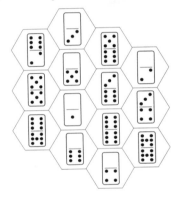

Keep up   Students play a game in which players randomly select seven dominoes from the facedown set and keep them hidden from their partners. The player with the 7–7 domino starts a game by taking one domino from his or her hand and placing it anywhere on the grid. Players take turns placing dominoes so that no two with consecutive values are in adjacent cells. The last player to place a domino is the winner.

Wrap-up   Key question for discussion or response in journals.

- Is there a best place on the grid to begin placing your dominoes? Explain.
- What strategy did you use to play the Keep-up game?

Follow-up   Have students play the game again, this time using the playing strategy or strategies that they considered the best. After they have played, ask them how well they were able to use their strategy, especially since their partner was now trying to use it too!

Next Hex

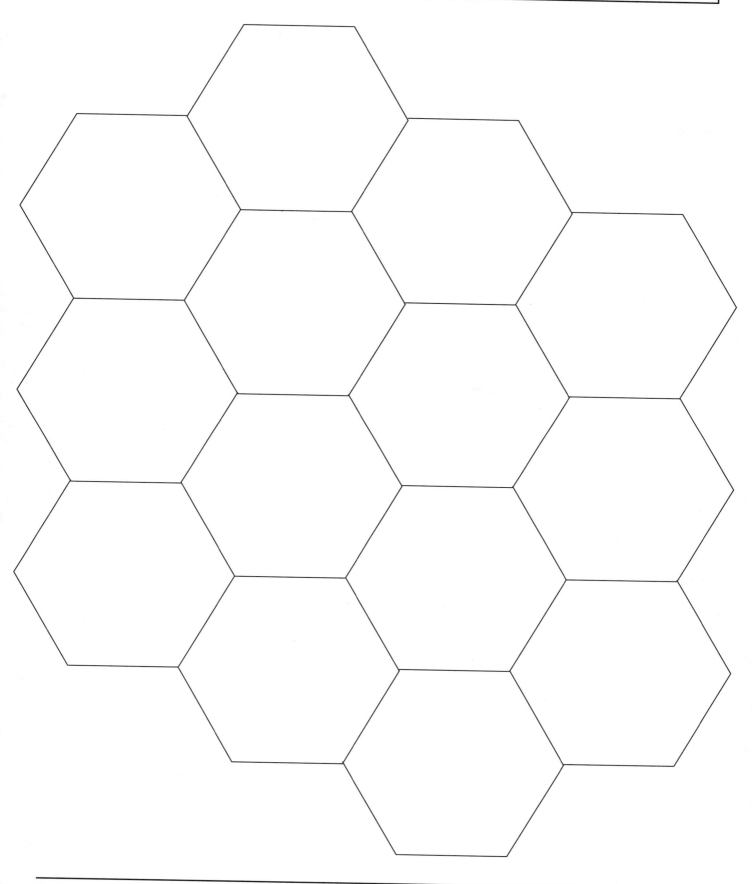

Butterfly 12

Task Students arrange seven dominoes having consecutive values in paths of three so that the value of each path—horizontally, vertically, and diagonally—is 12.

Setup Each pair of students uses a "Butterfly-12" mat and these dominoes taken from a double-nine set: 0–1, 0–2, 0–3, 0–4, 0–5, 0–6, 0–7. (The activity can also be done using these dominoes taken from a double-six set: 0–1, 0–2, 0–3, 0–4, 0–5, 0–6, 1–6.)

Start-up Students arrange the dominoes to form paths so that the sum of the values in each path is 12. You may want to suggest that students randomly place the dominoes, find the sums of the paths, and rearrange dominoes, as necessary, until each path has a sum of 12.

Discussion Be sure students realize that they must arrange the dominoes in five paths so that one domino is in three of the paths and each of the other dominoes is in just two. Some students might discover the advantage of starting with the domino at the center. Students should notice that if they cover the center domino, the other dominoes in each path have a sum of 8. For students having difficulty placing dominoes, it may help to have them arrange the seven in order of value, from the one with the least value (0–1) to the one with the greatest (0–7 or 1–6).

One sample solution:

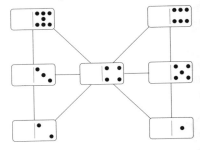

Keep up Challenge students to arrange these dominoes so that the paths have sums of 18: 0–0, 1–1, 2–2, 3–3, 4–4, 5–5, 6–6. Encourage them to think about how they may use what they learned from making paths of 12 to help them make paths of 18.

Wrap-up Key questions for discussion or response in journals.

- Is there a best place on the mat to begin placing your dominoes? Explain.
- If you cover the center domino, what do you notice about the remaining pairs?

Follow-up Have students make paths with sums of 36 using these dominoes: 0–9, 5–5, 5–6, 6–6, 6–7, 7–7, 7–8. You may have them also make paths with sums of 21 using these dominoes: 1–3, 1–4, 1–5, 1–6, 1–7, 1–8, 1–9.

Butterfly 12

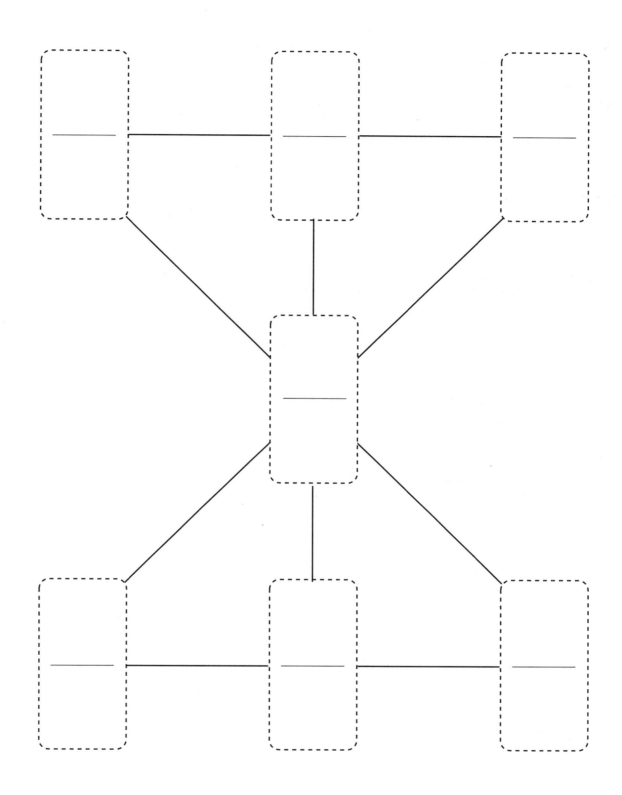

Math Activities with **DOMINOES**
© ETA/Cuisenaire®

Wheeling Around

Task   Students arrange eleven dominoes having consecutive values in paths of three so that the value of each path is 11.

Setup   Each pair of students uses a "Wheeling Around" mat and these dominoes taken from a double-nine set: 0–0, 0–1, 0–2, 0–3, 0–4, 0–5, 0–6, 0–7, 0–8, 0–9, 5–5. (The activity can also be worked using these dominoes taken from a double-six set: 0–0, 0–1, 0–2, 0–3, 0–4, 0–5, 0–6, 1–6, 2–6, 3–6, 4–6.)

Start-up   Students arrange the dominoes in a circular formation on the mat with one domino at the center, or Hub, so that the sum of the values of the three dominoes in each path is 11.

Discussion   Students should first notice the need to find five groups of three different addends, each with a sum of 11. After a while, they will probably conclude that the 0–0 domino should be at the Hub. Then they should determine that they need to find five pairs of dominoes, each with values that add up to 11. For students having difficulty placing dominoes, it may help to have them arrange the dominoes in order of value, from the one with the least value (0–0) to the one with the greatest (5–5 or 4–6).

One sample solution (double-nine dominoes):

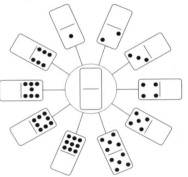

Keep up   Challenge students to arrange these dominoes so that the paths have sums of 23: 0–4, 0–5, 0–6, 0–7, 0–8, 0–9, 1–9, 2–9, 3–9, 4–9, 5–9. Encourage them to think about how they might use what they learned from making paths of 11 to help them make paths of 23.

One sample solution (double-six dominoes):

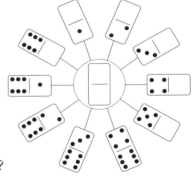

Wrap-up   Key questions for discussion or response in journals:

*   For any set of eleven dominoes whose values are consecutive, how can you easily find the Hub domino?

*   If you know the Hub domino for any set of eleven dominoes with consecutive values, how can you easily find equal values for the dominoes in the paths?

Follow-up   Have students arrange their own sets of dominoes with consecutive values in paths on mats of their own design.

Wheeling Around

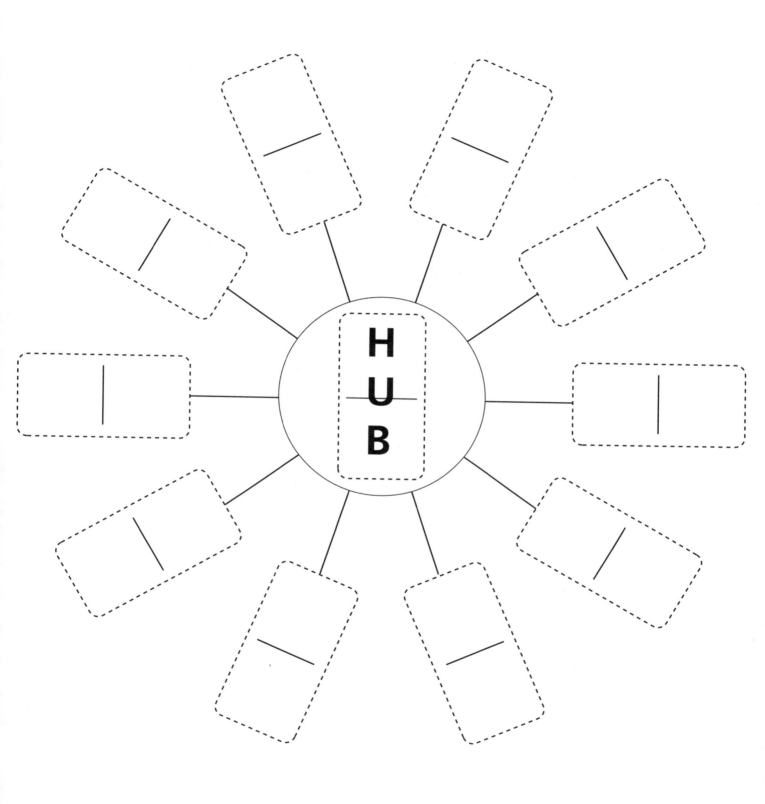

Task   Students arrange nine dominoes to form specific number sentences that involve all four arithmetical operations (addition, subtraction, multiplication, and division).

Setup   Students work independently using a "To Tell the Truth" mat and these dominoes taken from a double-six set: 0–1, 0–2, 0–3, 0–4, 0–5, 0–6, 3–4, 4–4, 3–6.

Start-up Students examine the mat and recognize that it provides outlines for four domino number sentences (three that are read horizontally and one that is read vertically). They arrange the dominoes between the symbols on the mat so that the values of the dominoes and the symbols together form true number sentences.

Discussion Students should notice that since the division sentence has the fewest possible solutions it might be advantageous to begin placing dominoes here. (The choices for solutions are limited because each value—from 1 to 9—occurs just once.)

Solution:

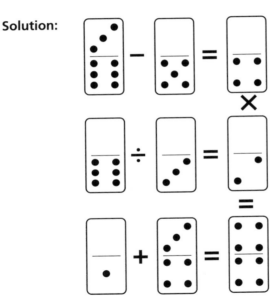

Wrap-up   Key questions for discussion or response in journals:

- In which number sentence did you start placing dominoes? Why?
- Do you think there is more than one solution to this problem? Explain.

Follow-up Have students choose any nine dominoes from a set and try to arrange them on a sheet of paper to make cross–number sentences. Tell them to record the operational symbols they used to create their own mats. Then have them exchange mats and challenge one another to find the dominoes that form true number sentences on their mats.

To Tell the Truth

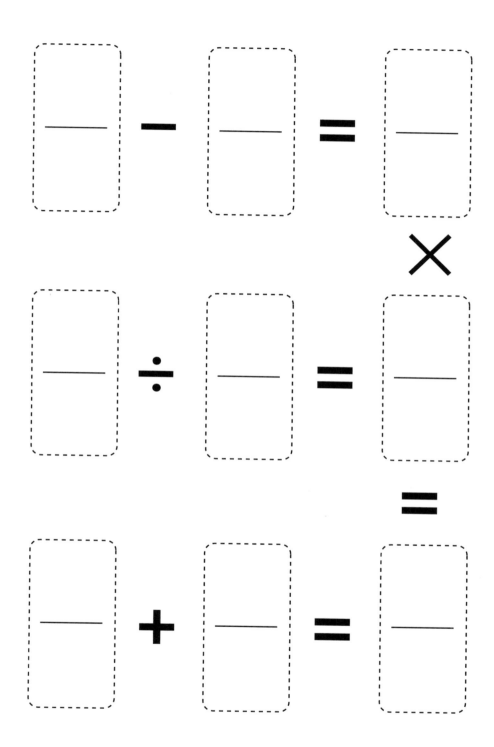

Task   Students use groups of four dominoes to form seven square arrangements so that for each square, the sum of the values of the faces on each side is the same.

Setup   Students work independently using one set of double-six dominoes and a "Quartets" mat. (A set of double-nine dominoes is needed for the Follow-up activity.)

Start-up   Model how to form a square "quartet" of four dominoes, how to find the sum of the three faces comprising each side of the square, and how to record the quartet.

Students manipulate dominoes to form as many quartets as they can. They check the sum of each side of a quartet as they find it. The dominoes used for one quartet are set aside and are not reused for other quartets.

Discussion   Students will use trial and error to make quartets and find the sums. As a quartet is made and recorded, students should slide it off the mat, keeping it in intact for future reference. Finding a complete set of seven quartets is a great challenge. Most students will find five or six quartets, then be unable to make others from the remaining dominoes. They may then look back at previously made quartets and decide to rearrange some of them in an attempt to find all seven.

One sample solution:

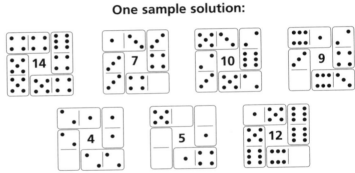

Keep up   Post students' solution sets. Students who have made and recorded all seven quartets may share their work by naming the dominoes they used for one or two quartets and challenging the class to find the quartets they made from the remaining dominoes.

Wrap-up   Key questions for discussion or response in journals:

- What is the least possible sum for a side of a quartet? What is the greatest possible sum? Explain.
- Is it possible to use more than one double in the same quartet? Explain.

Follow-up   Students can use 52 of the 55 dominoes from a double-nine set to make *thirteen* quartets in many ways.

Quartets

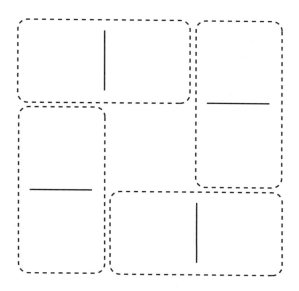

✂ -

Recording Sheet

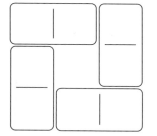 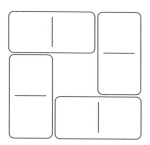 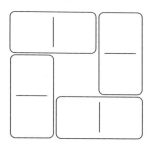 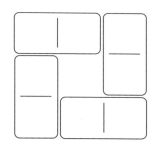

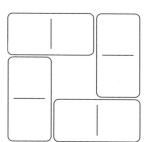 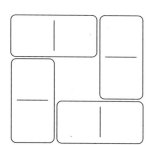 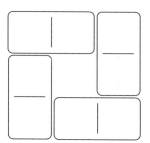

Task Students arrange a set of dominoes in a star formation so that the dominoes within each ray have matching faces and a sum of 21.

Setup Each pair of students uses one set of double-six dominoes and a "Starry 21" mat.

Start-up Identify the parts of the star formation as "rays." Explain that students should arrange the complete set of dominoes as shown on the mat so that in each ray the touching faces match and the sum of the values is 21.

Discussion Students should first notice that the star formation is made up of eight rays—four 3-domino rays and four 4-domino rays—each of which must have a sum of 21. Some students will use trial and error to arrive at a solution. Others may devise a strategic approach. One such approach might be to first arrange the complete set on the mat so each ray has a sum of 21, then to exchange the dominoes within each ray, as necessary, in order to find matching faces.

One sample solution:

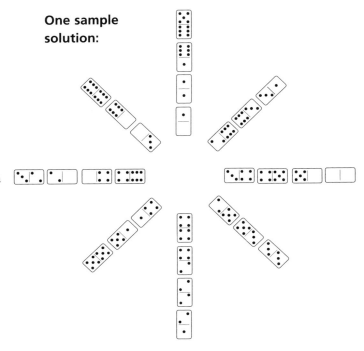

Keep up You may want to further challenge students to arrange the dominoes so that the faces of the innermost dominoes, those closest to the center of the circle, have pips representing at least one of each number from 0 to 6.

Wrap-up Key questions for discussion or response in journals:

- Did you use a strategy to start this activity? If so, how good was your strategy?
- How did you know where to place the dominoes with the least and greatest values?

Follow-up Have students position the dominoes in the star formation so that the innermost *and* the outermost faces both have pips representing at least one of each number from 0 to 6. Encourage students to design their own domino star formations with rays of differing lengths.

Starry 21

Septets

Task   Students use groups of seven dominoes to form four rectangles, each with adjacent matching faces and a sum of 42.

Setup   Students work independently or with a partner using one set of double-six dominoes and a "Septets" mat.

Start-up   Model how to form a rectangular "septet" of seven dominoes, how to find the sum of all the pips in the rectangle, and how to record the septet.

Students manipulate dominoes to form as many septets as they can. They check the sum of the pips in a septet as they find it. The dominoes used for one septet are set aside and not reused for other septets.

Discussion   Students will use trial and error to make septets and find sums equal to 42. Some students may first realize that since there are eight domino faces in the set, each with the numbers from 0 to 6, there must be two faces with the same number in each rectangle. Others may realize that wherever they use doubles, they must have two additional faces to match the two ends. Still others may first make a rectangle with dominoes whose pips add up to 42 and then attempt to manipulate the faces so that they match. Students may realize that, once they make three rectangles—each with a sum of 42—the dominoes that remain will have a sum of 42.

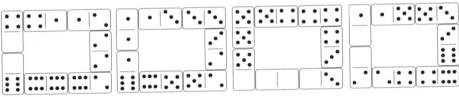

Sample solution:

Keep up   Post students' solution sets. Students who have made and recorded all four septets may share their work by naming the dominoes they used for one septet, and challenging the class to find the septets they made from the remaining dominoes.

Wrap-up   Key questions for discussion or response in journals:

- Why do some rectangles have four of one face and others have none of that face?
- Why must the sum of the values of the seven dominoes equal 42?

Follow-up   Challenge students to form squares having *four* dominoes, *six* dominoes, *eight* dominoes, and *ten* dominoes so that each has a sum of 42 and all touching faces match.

Solution:

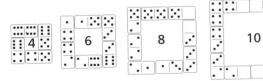

Septets

Logic

Math Activities with **DOMINOES**
© ETA/Cuisenaire®

41

What's Your Position?

Task
Students read clues that identify four dominoes with given numerical values. Then they order the dominoes by following the clues.

Setup
Each pair of students uses one set of double-six dominoes, "Clue Cards," and a "What's Your Position?" mat. (A set of double-nine dominoes may be used for the Follow-up.)

Start-up
Have students select the four dominoes shown on Clue Card A (1–1, 0–3, 3–6, and 4–6) and place the dominoes in front of them. Direct them to read along as you read the three clues aloud. Be sure students understand that they are to determine which dominoes belong in each position, from first to fourth, by lining up the dominoes in order on the mat, from left to right.

Discussion
Students should realize that since the 1–1 and 4–6 dominoes have even-numbered values, these dominoes should be in either of the first and fourth positions. Then, because the 0–3 and 3–6 dominoes have values that are multiples of 3 (and because the positions of the other two dominoes have already been identified), these dominoes should be in either the second or third positions. Finally, in order to satisfy the last clue, which calls for differences of 1 between the first two dominoes and the last two, students should realize that the 0–3 domino must be next to the 1–1 and the 3–6 domino must be next to the 4–6.

Keep up
Have students work together to determine the correct order of the dominoes for the other clue cards.

Possible solutions:

Clue Card	1st	2nd	3rd	4th
A	4–6	3–6	0–3	1–1
B	3–3	2–6	1–4	2–5
C	1–2	5–6	2–2	2–4
D	6–6	0–2	1–3	1–2

Wrap-up
Key questions for discussion or response in journals.

- Is it always best to start with the first clue? Explain.

- Which clues were easiest to follow? Which were most difficult? Explain.

Follow-up
Have pairs of students choose four or more dominoes, extend the mat if they choose more than four, and write their own set of clues. Have pairs exchange clue cards and challenge one another to order the dominoes correctly according to the clues.

Logic

Clue Cards

Clue Card A

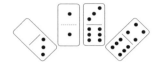

Clue 1: The values of the dominoes in the 1st and 4th positions are even.

Clue 2: The 2nd and 3rd dominoes have values that are multiples of 3.

Clue 3: The difference between the values of the first two dominoes and the last two dominoes is 1.

Clue Card B

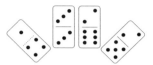

Clue 1: The dominoes that have consecutive values are not next to one another.

Clue 2: The sum of the values of the 2nd and 3rd dominoes is equal to the sum of the values of the 1st and 4th dominoes.

Clue 3: The difference between the faces of the domino in the 3rd position is the same as the difference between the faces of the domino in the 4th position.

Clue Card C

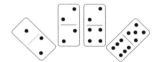

Clue 1: The dominoes in the 1st and 4th positions have values that are triangular numbers.

Clue 2: The dominoes whose values are prime are in the 1st and 2nd positions.

Clue Card D

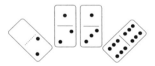

Clue 1: The dominoes in the 1st and 3rd positions have values that are multiples of 4.

Clue 2: The sum of the values of the first two dominoes is twice the sum of the values of the last two dominoes.

What's Your Position?

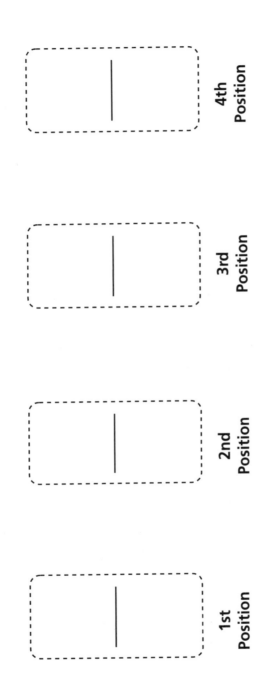

1st
Position

2nd
Position

3rd
Position

4th
Position

Patterns

Students observe and compare dominoes, look for similarities and differences among their attributes, and describe the numerical or geometric patterns that the attributes reflect. They use patterns and relationships to analyze mathematical situations in domino sequences, on a coordinate plane, and on a number chart.

Task   Students play a game in which they find three dominoes having common attributes to form a "trio."

Setup   A small group of students uses one set of double-six dominoes placed in a paper bag. (A set of double-nine dominoes may be used for a Follow-Up game.)

Start-up   The group picks seven dominoes from the bag without looking and places them face up on the playing area. Players each try to find three dominoes that together form a "trio." Dominoes form a trio if, when held vertically and lined up in a row,

    1. the three top faces have the same value or form a pattern;

    2. the three bottom faces have the same value or form a pattern.

Whoever first sees three dominoes that meet these conditions calls out "Trio." He or she puts the three together, describes the values and/or patterns, and takes the dominoes. Players pick three new dominoes, put them with the remaining four, and the play continues.

If no new trios are found during the game, an additional domino is picked and players try to use it to form a trio. If necessary, up to two more dominoes may be picked, but if still no trios are found, all the dominoes are returned to the bag and seven new ones are picked. When no more dominoes are left to be picked or when no more trios can be made, the player with the most dominoes wins.

Discussion   You may want to model examples of trios. Be sure students understand that when the three dominoes are lined up, the upper faces and/or the lower faces may be the same or they may form a pattern either from left to right or from right to left.

Sample trios:

Top: same values
Bottom: forms pattern

Top: forms pattern
Bottom: forms pattern

Top: forms pattern
Bottom: same values

Top and Bottom: form same pattern

Keep up   Have students record their trios then spill the remaining dominoes from the bag and choose the ones they need to extend the patterns.

Wrap-up   Key questions for discussion or response in journals:

- What was your strategy for finding trios?
- What type of pattern was easiest to find? Why?

Follow-up   Students may play the game with two sets of dominoes and try to find three, four, or five dominoes to form patterns. Students may also play the game using a set of double-nine dominoes. They would begin this game by picking *nine* dominoes for the playing area.

Task   Students play a game in which they find three dominoes having common attributes to form a "tri-color trio."

Setup   A small group of students uses three sets of double-six dominoes of different colors placed in a paper bag.

Start-up   The group picks nine dominoes from the bag without looking and places them face up on the playing area. Players each try to find three dominoes that together form a "tri-color trio." Dominoes form this kind of trio if, when held vertically and lined up in a row,

1. the three top faces have the same value or form a pattern;
2. the three bottom faces have the same value or form a pattern;
3. the dominoes are either all the same color or are of three different colors.

Whoever first sees three dominoes that meet all these conditions calls out "Trio." He or she puts the three together, describes the values and/or patterns and colors, then takes the dominoes. Players pick three new dominoes, put them with the six already on the playing area, and the play continues.

If no new tri-color trios are found, an additional domino is picked and players try to use it to form a tri-color trio. If necessary, up to two more dominoes may be picked, but if still no tri-color trios are found, the group returns all the dominoes to the bag and picks nine new ones. When no more dominoes are left to be picked or when no more tri-color trios can be made, the player with the most dominoes wins.

Discussion   You may want to model examples of tri-color trios and play a demonstration game. Students commonly make the error of trying to form a tri-color trio with two dominoes of the same color and one domino of another color. Some students may have difficulty focusing on several attributes at once. If so, they may need more experience playing "Trio," a similar game in which they use a single set of dominoes.

Sample tri-color trios:

Red Green Black

Red Green Black

Red Red Red

Keep up   Have students record their tri-color trios, spill the remaining dominoes from the bag, and choose the ones they need to extend the patterns.

Wrap-up   Key questions for discussion or response in journals:
- What was your strategy for finding tri-color trios?
- What type of pattern was easiest to find? Why?

Follow-up   Have pairs of students play a rummylike game. Players each pick six dominoes for their hands. One domino is turned over as the "Starter." A player picks a domino, either the Starter or a domino from the pile, and discards one domino. Players take turns picking and discarding. The first player to get two tri-color trios wins.

Finish It!

Task   Students use dominoes to identify, continue, and create patterns.

Setup   Students work independently or with a partner using one set of double-six or double-nine dominoes and a "Finish It!" mat. You may choose instead to cut several mats apart and distribute individual patterns for students to work on and exchange later.

Start-up   Students examine one of the patterns and copy it with dominoes. Be sure they understand that the blank lines tell how many more dominoes they must find to finish the pattern. Point out that an ellipsis (three dots) means that the pattern may be continued even further.

When students feel they have correctly completed their patterns, have them compare their work, resolve any differences, and then record the patterns.

Discussion Encourage students to look for alternative ways in which to complete each pattern. Any solution that is mathematically and/or logically valid should be accepted. Students should discover that patterns may be determined by the numerical order of the domino suit, by a numerical sequence made up of doubles (or other multiples), and by the orientation of the dominoes.

Sample solutions:

A   **rule: add 1 to top and bottom faces**

B   **rule: even numbered sums that increase by 2**

E   **rule: sums of 6, 5, 6, 5, 6, 5, 6**

G   **rule: alternate top to bottom to top: 0, 1, 2, 3, 4, 5, 6, ...**
    alternate bottom to top: 0, 0, 1, 1, 2, 2, 3, ...

Keep up   Encourage students to create their own six- or seven-domino patterns. Then have them remove two or three dominoes, leaving gaps to show their positions, and return these dominoes to the pile. Students should challenge one another to find the missing dominoes to complete their patterns.

Wrap-up   Key questions for discussion or response in journals:

- How did you decide which domino came next in a pattern?
- How did you go about creating your own pattern?
- Did you complete the pattern in exactly the same way as the student who created it? Explain.

Follow-up   Have students create more sophisticated patterns by positioning dominoes horizontally and reading them as two-digit numbers or by positioning them vertically and reading them as fractions. After students record these patterns, challenge them to continue the patterns with numbers that extend beyond those in the domino set.

Finish It!

A

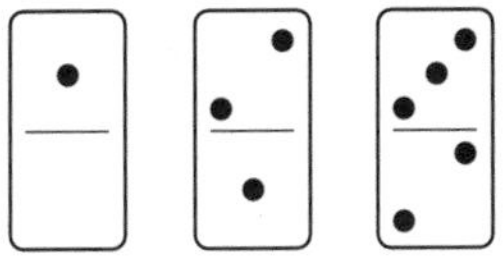

B

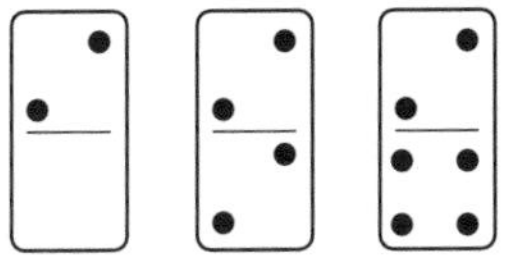

C

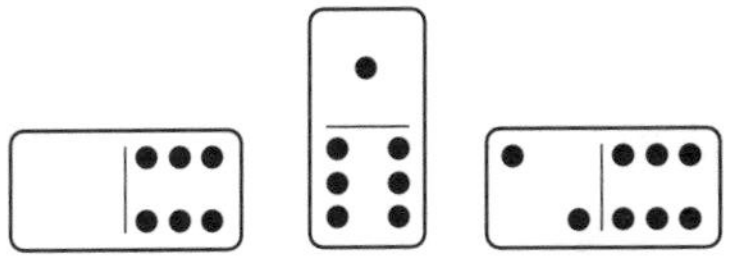

D

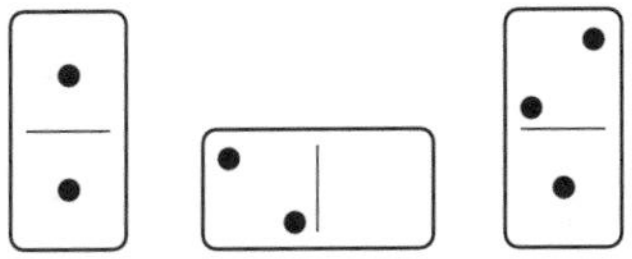

E

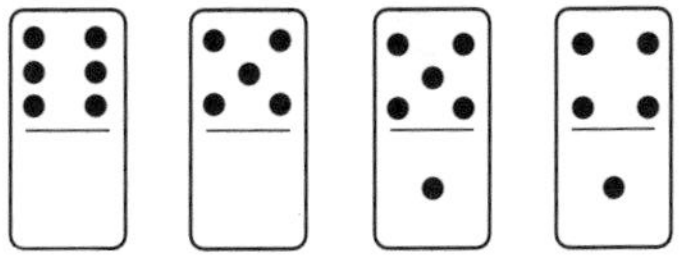

F

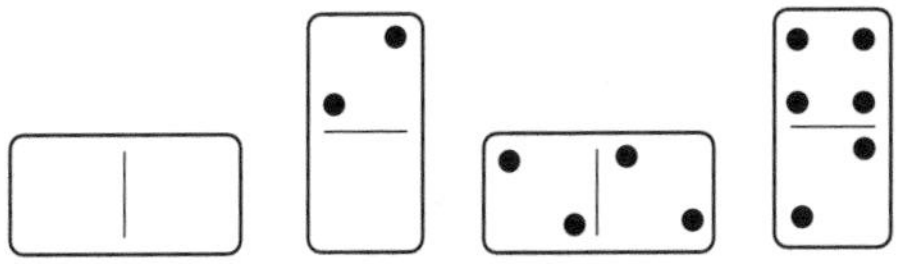

G

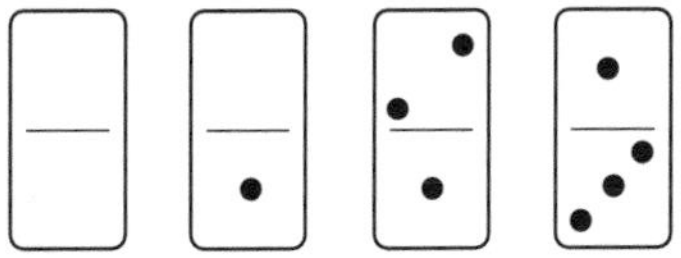

H Design your own domino pattern here.

Math Activities with **DOMINOES**

© ETA/Cuisenaire®

Get Coordinated!

Task   Students play a game in which they pick dominoes, read them as ordered pairs, and locate and cover corresponding points on a coordinate grid. Their goal is to be the first to cover four points in a row.

Setup   A pair of students uses one set of double-six dominoes placed in a paper bag and a "Coordinate Grid–(6, 6)" mat. Each student has 20 transparent chips of one color or a crayon or marker of one color. (A set of double-nine dominoes and a "Coordinate Grid–(9, 9)" mat is needed for the Follow-up game.)

Start-up   Hold up a domino horizontally and say that the faces may be read as an ordered pair. Use the 3–4 domino, for example, explaining that when it is held one way, it may be read as the ordered pair (3, 4) and when turned around, it may be read as (4, 3). Describe how each ordered pair relates to a point on the coordinate grid. Be sure students understand that the first number in an ordered pair tells how far to count *across* the grid from 0; the second number tells how far to count *up* from 0.

Students take turns picking a domino from the bag and deciding which way to read it as an ordered pair. They place a chip on the point named by the ordered pair and return the domino to the bag. If one of the coordinate points for a domino has already been covered, the player must cover the other point. Once both coordinates have been covered, players set the domino aside and do not return it to the bag. The first player to get four chips of his or her color in an uninterrupted row—horizontally, vertically, or diagonally—is the winner.

Discussion   As students play, they may alternately play to win and play to block. (Sometimes, students who are focused on blocking do not notice when they get four in a row!) As with other tic-tac-toe-like games, there may be a draw if all points have been covered and no one has gotten four in a row.

Keep up   Have students play the game again. This time, instead of having them play to cover points in a row, have them play to cover four points that when connected form a square or other predetermined shape.

Wrap-up   Key questions for discussion or response in journals:

- What kind of domino represents a single point whichever way it is read?

- As you played, did you try harder to get four in a row or to keep your partner from getting four in a row? Explain.

Follow-up   Have students play the game on the Coordinate Grid–(9, 9) mat using a set of double-nine dominoes. The winner of this game is the player who is the first to cover *five* points in a row.

Coordinate Grid–(6, 6)

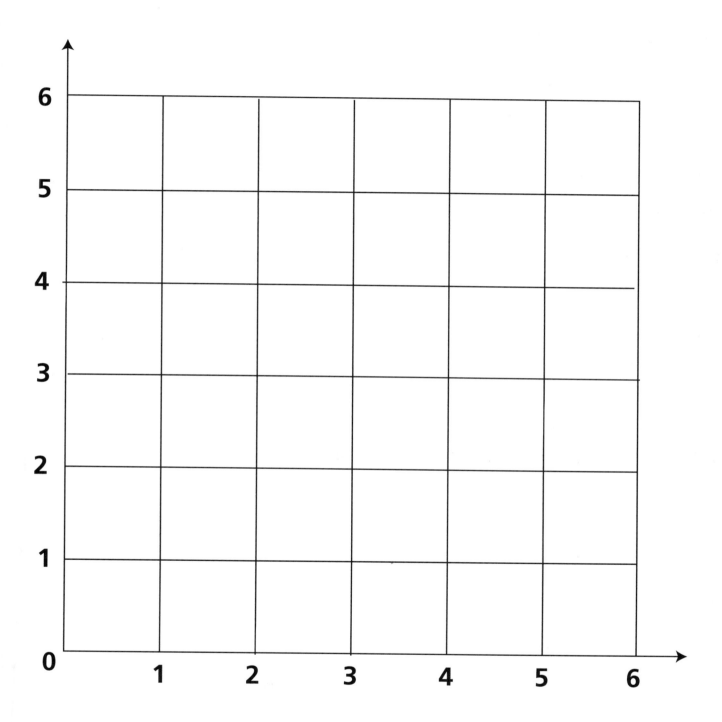

Coordinate Grid–(9, 9)

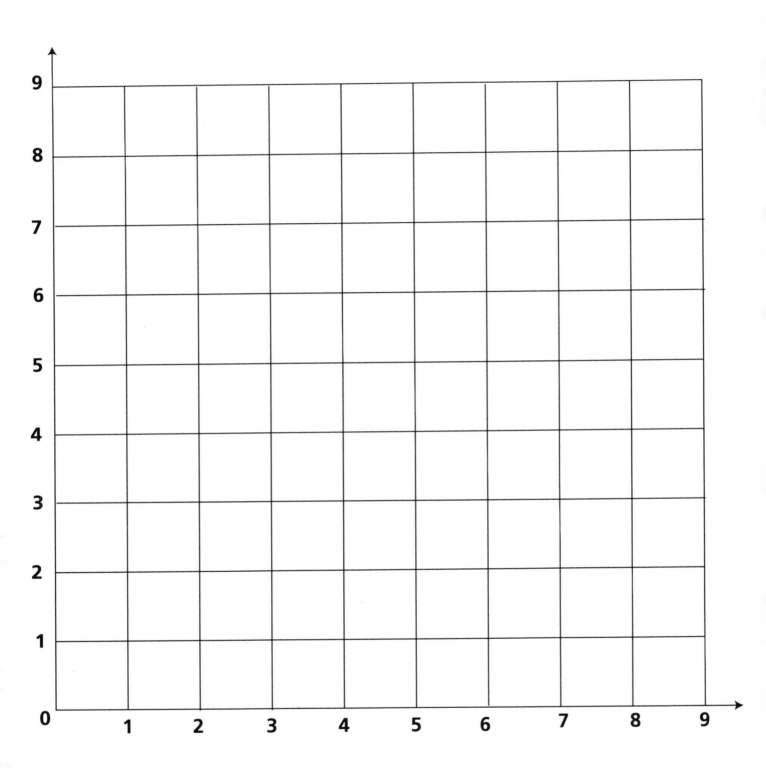

Point Search

Task   Students play a game in which one player secretly marks points on a coordinate grid. The other player picks dominoes and reads them as ordered pairs, trying to locate the hidden points.

Setup   A pair of students uses one set of double-six dominoes placed in a paper bag and a stack of books or a file folder to serve as a barrier between them. Each student has 20 transparent chips of one color or a crayon or marker of one color and a "Coordinate Grid–(6, 6)" mat. (A set of double-nine dominoes and a "Coordinate Grid–(9, 9)" mat may be used for a Follow-up game.)

Start-up   Hold up a domino horizontally and explain that the faces may be read as an ordered pair. Use the 3–4 domino, for example, and point out that when held one way, it may be read as the ordered pair (3, 4) and when turned around, it may be read as (4, 3). Describe how each ordered pair relates to a point on the coordinate grid. Be sure students understand that the first number in an ordered pair tells how far to count *across* the grid from 0; the second number tells how far to count *up* from 0.

Players decide on roles—Plotter and Searcher. The Plotter marks four coordinate points on his or her grid. The Searcher, trying to guess the four points, starts by picking a domino from the bag and naming one of its ordered pairs. The Plotter responds with "Hit" if the ordered pair corresponds to one of his or her points or "Miss" if it doesn't. The domino is returned to the bag. After naming 20 ordered pairs, the total number of hits is recorded. Players exchange roles and play again. The winner is the player who has the most hits after two rounds.

Discussion   Searchers can keep track of the ordered pairs they name by marking them on their own grids, perhaps by circling each hit and crossing out each miss. Alternatively, Searchers might want to record the hits and misses in a table such as the one started at the right.

Hit	Miss
5, 1	
	3, 4
	2, 0
0, 2	

Wrap-up   Key questions for discussion or response in journals:

- Which role did you like better, Plotter or Searcher? Why?
- How would your strategy change if you could see the dominoes you pick?

Follow-up   The object of another version of the game is to locate a polygon. The Plotter draws a polygon on the grid by connecting coordinate points. When the Searcher names an ordered pair, the Plotter responds with "Hit," "Miss," or "Vertex Hit." (A *hit* is any point on a side or within the polygon. A *miss* is a point outside the polygon. A *vertex hit* is a point at the vertex of the polygon.) To increase the difficulty of this game, Plotters may draw more than one polygon on a single grid.

Have students play either of the games on a Coordinate Grid–(9, 9) mat using a set of double-nine dominoes.

Spot Check

Task   Students play a game in which they pick dominoes, identify them as two-digit numbers, and mark the numbers on a grid. Their goal is to be the first to form a numerical sequence or pattern.

Setup   Each pair of students uses one set of double-six dominoes placed in a paper bag and a "Double-6 Spot Check" mat. Each student needs 20 transparent chips of one color or a crayon or marker of one color. (A set of double-nine dominoes and a "Double-9 Spot Check" mat is needed for the Follow-up game.)

Start-up   You may want to model several number patterns on the Double-6 Spot Check grid by covering numbers with chips or by circling them. Patterns do not have to be made up of adjacent numbers only. They may reflect consecutive even or odd numbers or multiples of numbers. A pattern may be diagonal, like this one, which reflects repeated additions of eleven—1, 12, 23—or like this one, which reflects repeated additions of twelve—1, 13, 25.

Players take turns picking a domino from the bag, holding it horizontally, and reading it as a two-digit number. (If, for example, the 4–2 domino is picked it could be read as 42 or as 24.) Then players place a chip on the number they choose on their grid or else they circle the number. Picked dominoes, except for doubles and those that have been read in both ways, are returned to the bag. The first player to complete a three-number pattern or sequence is the winner.

Discussion Recognizing patterns and finding relationships between numbers is important to the understanding of math concepts. Some students will identify only familiar patterns, whereas others will identify more complex numerical relationships. Some students may notice the visual configuration of patterns on the grid. Others may analyze the position of previously placed chips, determine how these could be used to begin a pattern, and project which numbers they would need to pick in order to finish the pattern. More insightful students may use blocking strategies, trying to keep their opponents from finishing their patterns.

Keep up   Students may use arrow statements such as the following to record the numbers and the movement on the grid for their winning patterns: 11↓→ → 23↓→ → 35

Wrap-up   Key questions for discussion or response in journals:
* Why were some patterns more difficult to continue than others?
* Which patterns were easiest to spot?
* Did you use a strategy to complete a pattern? What was it? Was it successful?

Follow-up Have students play the game using a set of double-nine dominoes and the 0 to 99 grid. The goal of this game is to complete a *four*-number sequence or pattern.

Double-6 Spot Check

0	1	2	3	4	5	6
10	11	12	13	14	15	16
20	21	22	23	24	25	26
30	31	32	33	34	35	36
40	41	42	43	44	45	46
50	51	52	53	54	55	56
60	61	62	63	64	65	66

Math Activities with DOMINOES
© ETA/Cuisenaire®

Double-9 Spot Check

0	1	2	3	4	5	6	7	8	9
10	11	12	13	14	15	16	17	18	19
20	21	22	23	24	25	26	27	28	29
30	31	32	33	34	35	36	37	38	39
40	41	42	43	44	45	46	47	48	49
50	51	52	53	54	55	56	57	58	59
60	61	62	63	64	65	66	67	68	69
70	71	72	73	74	75	76	77	78	79
80	81	82	83	84	85	86	87	88	89
90	91	92	93	94	95	96	97	98	99

Number Theory

Students compare and generalize as they identify number properties and discover relationships among integers. They work with prime and composite numbers, multiples and factors, and odd and even numbers as they compete in a variety of strategy games and activities.

Task   Students play a game in which they try to reach a target number by performing arithmetical operations on the numerical values of domino faces.

Setup   A small group of students uses a set of double-six dominoes and a set of "0 to 12 Number Cards." Each student has 15 chips of one color and a "Target Squares" mat.

Start-up   The dominoes are placed face down on the playing area and the number cards are mixed and stacked face down. Students each pick three dominoes. They turn over one card to show the target number for that round. Students mentally add, subtract, multiply, or divide the two faces of each of their dominoes, trying to get a result that equals the target number. They tell their number sentences to the group and/or record them. For each completed number sentence, players place one chip on the target number on their mat. In any one round, players may put down as many as three chips or no chips at all. The dominoes are all returned to the pile and another number card is turned over to show the next target number. After five rounds, the player who has placed the most chips on the board is the winner.

Discussion   As the round progresses, some students become aware of which dominoes are still to be played. This awareness is helpful in choosing operations that have whole-number solutions. Students should realize that, for example, in order to divide, one domino face has to be a multiple of the other.

One sample hand for the target number 1:

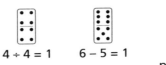

$4 \div 4 = 1$ $6 - 5 = 1$ not possible

(Two chips are placed in the 1 box on the mat.)

Keep up   Have students sort the dominoes for a given target number into two groups—those that can equal the target number and those that cannot.

Wrap-up   Key questions for discussion or response in journals:
- How did you decide which operation to use on the faces of a domino?
- For which target number was it most difficult to complete number sentences?

Follow-up   An even more challenging game can be played by using the faces of more than one domino in a hand to reach a target number. For example, all the dominoes in the hand shown above can be used to form a single number sentence for the target number 1.

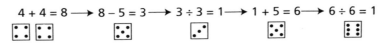

$4 + 4 = 8 \longrightarrow 8 - 5 = 3 \longrightarrow 3 \div 3 = 1 \longrightarrow 1 + 5 = 6 \longrightarrow 6 \div 6 = 1$

0 to 12 Number Cards

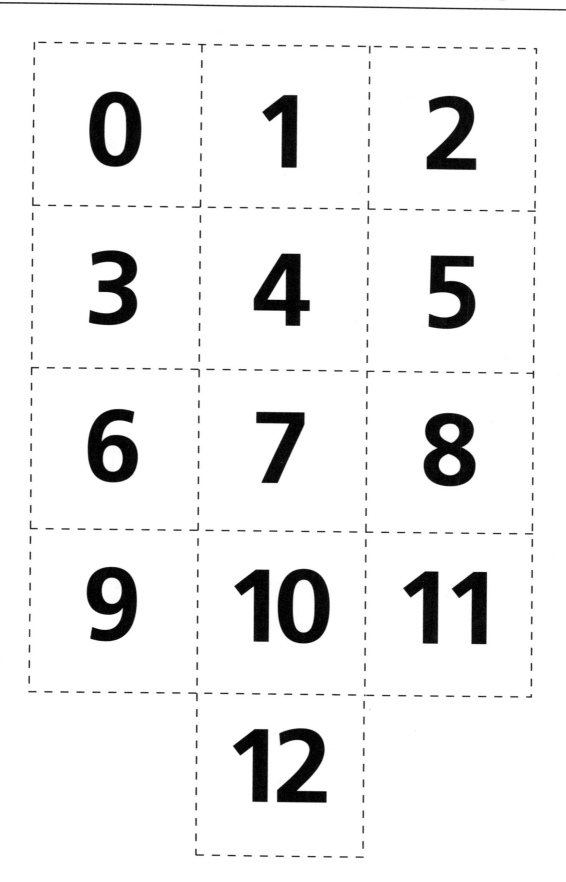

Target Squares

0	1	2
3	4	
5	6	7
8	9	
10	11	12

Task   Students play a game in which they try to score either the least or the greatest number of points by performing arithmetical operations on the numerical values of domino faces.

Setup   Each pair of students uses a set of double-six dominoes. Each student has a "Target Squares" mat. (A set of double-nine dominoes is needed for the Follow-up game along with a "Target Squares" mat extended at the bottom to include the numbers 13 to 18.)

Start-up The dominoes are placed face down in the playing area. To begin a round, one student determines the goal—to get the least or the greatest number of points—by spinning a spinner, tossing a coin, or rolling a die. Players take turns picking until each has four dominoes. In a round for the *least* number of points, a player adds, subtracts, multiplies, or divides the two faces on a domino to identify the lowest possible whole-number value. The player places the domino on the mat in the box corresponding to that value, says the number sentence aloud, and/or records it. At the end of the round, each player finds his or her score. The player with the lowest score receives one point for the round. (In a round for the *greatest* number of points, the player with the highest score receives the point.) All dominoes are returned to the pile. To begin the next round, least or greatest is determined and play continues. The first player to earn 5 points is the winner.

Discussion Students should realize that, to get a *least* solution, they must subtract or divide; to get a *greatest* solution, they must add or multiply. As the round progresses, some students become aware of which dominoes remain to be played. This is helpful in choosing operations with whole-number solutions. Students should notice that, for example, in order to divide, one domino face must be a multiple of the other.

One sample round for "least":

Player 1	Player 2
$3 \times 0 = 0$	$6 \div 2 = 3$
$5 - 4 = 1$	$2 - 2 = 0$
$4 \div 2 = 2$	$5 - 1 = 4$
$6 \div 3 = \underline{2}$	$3 - 1 = \underline{2}$
Score: 5	Score: 9

Player 1 wins the round and gets 1 point.

Keep up Have students play the game again, this time without returning the dominoes used in a round to the playing area.

Wrap-up Key question for discussion or response in journals:

• Which operation was the most difficult to use? Why?

Follow-up Have students play the game using a double-nine set and a Target Squares mat extended to include the numbers 13 to 18. Students pick *six* dominoes. Encourage them to use multiple operations by having each student add a bonus point to his or her score for each operation used beyond the first.

One sample hand for "greatest":
Score: 53 + 2 bonus points = 55 points

$8 + 8 = 16$ $(4 - 0) + (3 \times 3) = 13$ $(5 \div 1) + (7 + 3) = 15$ $5 + 4 = 9$

Factor Force

Task   Students play a game in which they pick a domino and read it as a two-digit number. They either identify that number as being *prime* or they name its *factors*.

Setup   Each pair of students uses one set of double-six dominoes with the double-blank domino removed and a "Factor Force" mat. Each student has 20 transparent chips of one color.

Start-up   Before having students play this game, you may want to review how to read a domino as a two-digit number with the greater number of pips in the tens place. You may also want to review the meanings of the terms *prime* and *factor*.

Students take turns picking a domino from the facedown set and reading it as a two-digit number with the face having more pips in the tens place. They decide whether or not the number is prime. If it is prime, they record the pick by putting a chip on the mat on any square marked "Prime." If the number is not prime, they record the pick by putting a chip on any one of its factors. For example, someone who picks the 3–2 domino may put a chip on any one of the factors of 32. Picked dominoes are set aside. Once a chip is placed on the mat, it may not be moved. The first player to get three chips in a row wins. (If all the dominoes have been played and no one has gotten three in a row, the player with the most chips on the mat wins.)

Discussion Students may notice that the mat has fewer of some factors than of others. Early in a game, while students still have the choice of many factors to cover, they should cover those that are less likely to become a choice again. Some students may decide to play defensively, trying to block their opponents rather than "going for a win."

Keep up   Have students record the factors of the numbers they pick or tally the number of times the factors are chosen during the course of one game. Students can analyze and use the data to help them make choices in future games.

Wrap-up   Key questions for discussion or response in journals:

- How do you know whether or not a number is prime?
- Which factor of the nonprime numbers is it possible to get most often? Why?

Factor Force

Prime	4	5	Prime	21	32
2	8	10	3	7	Prime
3	Prime	15	6	13	2
6	16	2	11	20	9
31	11	Prime	30	5	33
Prime	10	22	4	Prime	2

Task Students sort dominoes according to a given rule and solve a probability problem.

Setup Each pair of students uses one set of double-six or double-nine dominoes and a "What Are the Odds?" recording mat.

Start-up Tell students that they will play a computerlike game against an imaginary computer. Then they will decide whether or not they and the computer had equally likely, or fair, chances of winning.

Students take turns picking dominoes from the set turned face down on the playing area. They record the domino and its value and decide whether the value is odd or even. If it is odd, the pair wins the pick. If it is even, the computer wins. (Be sure that students understand that zero should be considered as an even number and a multiple of every number.) Picked dominoes are set aside. Students should play and record the results of four games.

Discussion After students discuss their results, have them sort a complete set of dominoes into two groups—those with odd values and those with even values. Encourage them to look for patterns. They should discover that the sum of two odd faces is even, the sum of two even faces is even, and the sum of an odd face and an even face is odd. They may also discover that if they use a double-six set, the computer can win with sixteen dominoes but they can win with only twelve. So the odds of the computer winning are *16 to 12*. (Using a double-nine set, the odds of the computer winning are *30 to 25*.)

Keep up Have students play the game a few more times, each time changing the rule so that the computer wins according to one of the rules listed in the table at the right. After each game, students can determine the odds of winning using that rule.

| RULE | ODDS OF COMPUTER WINNING | |
Computer wins:	Double-6 Set	Double-9 Set
if the value of the domino is greater than 9.	4 to 24	25 to 30
if the product of the faces is even.	22 to 6*	40 to 15*
if the value of the domino is a multiple of 3.	10 to 18*	19 to 36*
if the value of the domino is a multiple of 3 or 5.	15 to 13*	27 to 28*

* Consider 0 to be an even number and a multiple of every number.

Wrap-up Key questions for discussion or response in journals:

- How can you predict the winner using the odd/even rule for winning?
- What could you do to make the game fair?

Follow-up Challenge students to write their own rule for a game of this kind that would be "fair" because it gives players equal chances of winning.

What Are the Odds?

For each pick, circle *odd* or *even*. Who wins?

Pick	Record Domino	Record Value	Circle Winner
1st			odd | even
2nd			odd | even
3rd			odd | even
4th			odd | even
5th			odd | even

Pick	Record Domino	Record Value	Circle Winner
1st			odd | even
2nd			odd | even
3rd			odd | even
4th			odd | even
5th			odd | even

Pick	Record Domino	Record Value	Circle Winner
1st			odd | even
2nd			odd | even
3rd			odd | even
4th			odd | even
5th			odd | even

Pick	Record Domino	Record Value	Circle Winner
1st			odd | even
2nd			odd | even
3rd			odd | even
4th			odd | even
5th			odd | even

Clip or Chip It

Task Students play a game in which they pick a domino and match either the value of one face or the sum of the faces to boxes on a game board.

Setup Each pair of students uses one set of double-six dominoes. Each student has a "Clip or Chip It" mat and 15 paper clips or plastic chips. (A double-nine set of dominoes may be used for the Keep-up activity.)

Start-up Students take turns picking a domino from the facedown set and placing one clip or chip on their mat in the box whose number matches either one of the domino faces or the sum of the faces. The domino is then set aside. When no more dominoes are left to be picked, the player who has placed clips in the most boxes wins. The game may also be played for several rounds, with students scoring a point for each box in which they place a clip. Then, whoever has the higher score after a predetermined number of rounds is the winner.

Discussion As students decide where to place a chip, they may consider which other dominoes can also be placed in the same box on the mat. Some students may notice that there are several dominoes for each of the values 5, 6, and 7 but only one domino each for the values 11 and 12.

Keep up Have students play the game in a small group using a double-nine set of dominoes and a game board extended to include the numbers 13 through 18.

Wrap-up Key questions for discussion or response in journals:

- What are some things you can do to win this game?
- How would it affect the game if the dominoes were returned to the pile so that they could be picked more than once?

Follow-up Have students play this game in reverse! Students start by placing their clips or chips on their game boards in any boxes they choose. (They may put one or more clips in a box or they may put no clips in a box.) Students take turns picking dominoes from the facedown set. For each pick, they remove just one clip from the box whose number matches either one of the domino faces or the sum of the faces. Once a domino is played, it is set aside. When no more dominoes are left to be picked, the player with the fewest clips on the game board wins. The game may also be played for several rounds, with students getting a point for each clip remaining. Then, whoever has the fewest points after a predetermined number of rounds is the winner.

Clip or Chip It

0	1	2	3	4	5	6
7	8	9	10	11	12

And the Difference is...

Task   Students identify similarities and differences among four dominoes.

Setup   A small group of students uses one set of double-six or double-nine dominoes and a sheet of paper folded into quarters.

Start-up   Before having groups proceed on their own, you may want to use one of the solution sets below to model the way in which four dominoes may be evaluated.

Each group chooses four dominoes from the faceup set. They place the dominoes on the paper, one to a quarter, so that they are either all vertical or all horizontal. They discuss how each domino is like the others and different from the others. When dominoes are placed vertically, students consider the individual faces—finding the sums, differences, quotients, products, and factors for a pair of faces and/or reading the dominoes as fractions. When dominoes are placed horizontally, students read them as two-digit numbers. Students record their findings.

Discussion   Encourage students to choose their four dominoes so that each is like the others in at least one way and different from the others in at least one way. As students describe likenesses and differences, they use mathematical language and communicate important ideas.

Two sample solution sets:

Domino A: Both faces are even numbers. As a fraction, it is equal to 1/2. It has the only denominator that is not 3.

Domino B: As a fraction, it has the only numerator that is greater than the denominator.

Domino C: Its value is less than the value of any other domino.

Domino D: It has the only odd-numbered sum that is also prime. The difference between its faces is not 2.

Domino A: Twenty-three is the only prime number. It is not a multiple of 3 or 5.

Domino B: The sum of its faces are even. Both faces are odd.

Domino C: It is the only multiple of 9.

Domino D: It is the only number with an even number (0) in the ones place.

Keep up   Students may read the dominoes in a different way. For example, if at first they read them as two-digit numbers, they may now read them as decimals or fractions. Have students discuss how this new way of reading the dominoes creates changes in likenesses and differences between the first way of reading them and the new way.

Wrap-up   Key question for discussion or response in journals:
*   Which domino would always be most different from the others no matter which group it was in? Explain.

Follow-up   Challenge students to design their own "And the Difference is..." problems using a double-nine set. This work may be placed in the math center for others to try.

Computation

Students estimate values, make calculations, and match dominoes as they apply mixed computation to a variety of puzzles and games. They demonstrate their understanding of whole numbers and proper and improper fractions by using dominoes to represent number sentences, by forming magic squares, and by finding equivalent fractions.

R.A.P.–It!

Task   Students play a game in which they *roll* dice (R.) to find a sum, *add on* (A.) to match the sum, then *place* (P.) a domino to continue a domino path.

Setup   A small group of students uses one set of double-six dominoes and a pair of dice. (A set of double-nine dominoes and three dice are needed for the Follow-up game.)

Start-up   Students each pick four dominoes from the facedown set. One domino from the set is turned over as the "Starter." Each player, in turn, rolls one or two dice and, if possible, places a face of one domino from his or her hand with either face of the Starter so that the sum of these two faces equals the number rolled. Players pick a domino from the set on every turn even if they are unable to place a domino. As play continues, dominoes are placed only at the ends of the path. The winner is the first player to run out of dominoes. If no one runs out, and no more dominoes can be played, the player with the fewest pips on his or her remaining dominoes wins.

Discussion   As the game progresses and a path is formed, you may want to have students point out the end faces that can be matched on their turn. Be sure students understand that their choice of rolling one die or both dice depends on the domino faces at the ends of the path and on the dominoes in their hands. For example, if the end faces show high numbers such as 5 or 6, the player should roll both dice to increase the likelihood of being able to place a domino. With experience, students will realize when to roll one die or both dice. It is to their advantage to play dominoes with the most pips as soon as they can because if no one goes out, the player with the fewest pips on his or her remaining dominoes wins.

Part of a sample game:
1. The 5–3 domino is turned over as the Starter.
2. The first player uses *one die* and rolls a 4, then puts his 1–6 domino next to the Starter to show that the two adjacent faces have the sum of 4.
3. The second player uses the *pair of dice*, rolls a 10, and then puts her 4–0 domino next to the first player's domino to show that the two adjacent faces have the sum of 10.

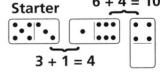

Wrap-up   Key questions for discussion or response in journal:

- On your turn, how did you decide whether to roll one die or the pair of dice?

- Which dominoes are best to play early in the game? Explain.

Follow-up   Have students play the game with a set of double-nine dominoes and *three* dice. Then have them compare this game with the double-six version.

Suit Yourself

Task   Students play a game in which use they use the dominoes in their hand to match the value or suit of a "Starter" domino.

Setup   A small group of students uses one set of double-six dominoes. (A set of double-nine dominoes is needed for the Follow-up game.)

Start-up   Students each pick five dominoes from the facedown set and stand them up so that only they can see them. One domino is turned over as the "Starter." Players take turns putting down a domino from their hand that has either the same value as the Starter or the same suit as one of its faces. Players who cannot match either the value or a suit keep picking dominoes from the facedown set until they get a match. Whoever puts down the domino with the least value takes all the dominoes for the round. (If two or more dominoes have the same value, the player who was first to put down a domino with that value takes them all.) A new Starter is turned over, and a new round begins. When no more dominoes can be played, the player who has taken the most dominoes wins.

Discussion   Students must understand that when they have a choice of dominoes to play, they should play the one with the least value. Encourage students to consider the effects of playing a particular domino on the hands of their opponents, the advantages of stockpiling a particular value or suit, the use of doubles to restrict options, and the probability of picking a domino in a suit with a given value.

One sample round:

Starter	Player 1's domino	Player 2's domino	Player 3's domino

Player 1 wins the round.

Wrap-up   Key question for discussion or response in journal:

- Which domino should you put down on your turn? Explain.

Follow-up   Have students play the game using a set of double-nine dominoes. They should start this game by each picking seven dominoes. After a few rounds, have students increase the difficulty of the game by playing so that they match suits alone and not values.

Thirty Sum-Thing

Task Students arrange nine double dominoes in paths in a square formation so that the sum of the values in each path—horizontally, vertically, and diagonally—is 30.

Setup Each pair of students uses a sheet of paper folded into nine boxes and these dominoes taken from a double-nine set: 1–1, 2–2, 3–3, 4–4, 5–5, 6–6, 7–7, 8–8, 9–9.

Start-up Students arrange the dominoes on the paper, one to a box, so that the sum of the values in each row, column, and diagonal path is 30.

Discussion Students should notice that each domino is part of either two or three paths. Some students might discover that there is an advantage to starting by positioning the dominoes that are in three paths.

One solution:

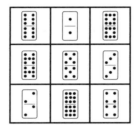

Keep up Challenge students to arrange these nine near-double dominoes in paths with sums of 27: 0–1, 1–2, 2–3, 3–4, 4–5, 5–6, 6–7, 7–8, 8–9. Encourage students to think about how they might use what they learned from making paths of 30 to help them make paths of 27.

One solution:

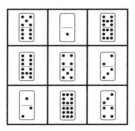

Wrap-up Key questions for discussion or response in journals:

- Is there a best place in which to put down the first domino?
- What do you notice about the two dominoes on opposite sides of the center domino in every path?

Follow-up Students can play a tic-tac-toe-like game. Have them use two sets of doubles, 1–1 through 9–9, taken from two sets of double-nine dominoes to form a path of 30. The first player to form one path—horizontally, vertically, *or* diagonally—is the winner.

Circle Sums

Task   Students place dominoes on a Venn diagram so that the sum of the values within each circle is equal to the sum of the values in each of the other circles.

Setup   Students work independently or with a partner using one set of double-six dominoes and a Venn diagram drawn on large sheets of paper or made from three pieces of string, each about 4 ft long and tied into a loop. (A fourth piece of 4-ft-long string is needed for the Follow-up activity.)

Start-up   You may want to model the activity using some of the zero-suit dominoes as shown at the right.

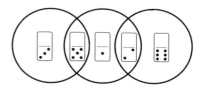

Students place the string loops on the playing area to form overlapping circles. They place one domino into each section of the diagram so the sum of the values in one circle equals the sum of the values in each of the other circles. Students may need a reminder that a domino in an intersection is lying in more than one circle. Students choose any dominoes from the entire set and try to find different solutions for each of the possible sums and record them.

Discussion   Students will discover that there are many ways to make various sums within the circles. At first, they may use trial and error to find a solution. Later, they may use information obtained from one solution to arrive at another.

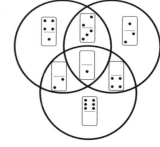

Keep up   After students have found a variety of solutions and have recorded the sums, have them look for patterns for getting sums. They might want to make Venn diagrams using other circle patterns such as the one at the right.

Wrap-up   Key questions for discussion or response in journals:

- Did you notice any patterns as you changed one solution to form another? Explain.

- For which sums did you find the most solutions?

Follow-up   Provide students with additional challenges by having them place dominoes with values from 1 to 7 into *four* intersecting circles. Then let them design their own circle patterns for dominoes that represent two-digit numbers, decimals, or fractions.

Sample Solution:

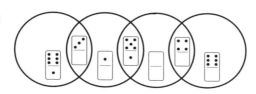

Computation   Math Activities with **DOMINOES**   **73**

In Disguise

Task   Students play a game in which they read a domino as a fraction and identify an equivalent fraction.

Setup   Each pair of students uses an "In Disguise" mat and one set of double-six dominoes with the zero-suit and the doubles removed. Each student has 10 chips of one color or a crayon or marker of one color.

Start-up   Hold up a domino vertically, pointing out that it can be read as a fraction with the face having fewer pips (the numerator) above the face with more pips (the denominator). Hold up the 3–4 domino, for example, and say that it can be read as "three-fourths." Then explain that the fractions on the mat are "in disguise" for the fractions that students will read.

Each student, in turn, picks a domino from the facedown set, reads it as a fraction, and tries to find an equivalent fraction on the mat. When one is found, the student covers it with a chip. Once a chip is placed on the mat, it may not be moved. After a domino is picked and played, it is set aside and cannot be reused. The first player to have chips in three squares that together form a right angle—two across and one up (or down) or two up (or down) and one across—is the winner.

Discussion Students will notice that there are several equivalent fractions on the mat for certain fractions they name, such as ½, but only one equivalent fraction for others, such as ⁵⁄₆. Depending on the dominoes they pick, students need to apply different skills and techniques. Some domino values may have to be expressed in lowest terms. Their equivalent values on the mat may also have to be expressed in lowest terms so that a match can be made. This provides opportunities for students to calculate mentally, although some students may need paper and pencil to do the calculations.

Keep up   Have students record the fractions they read and explain what they did to express each as an equivalent fraction in lowest terms.

Wrap-up   Key questions for discussion or response in journals:

- Which equivalent fractions were easiest to recognize? Why?

- What did you do to express a fraction in lowest terms?

Follow-up Students may work in pairs putting each domino described in the Setup on an equivalent fraction on the mat. Since there are fifteen dominoes and sixteen squares, one square will remain uncovered.

In Disguise

$\dfrac{5}{20}$	$\dfrac{9}{18}$	$\dfrac{12}{16}$	$\dfrac{3}{9}$
$\dfrac{12}{24}$	$\dfrac{3}{15}$	$\dfrac{4}{24}$	$\dfrac{4}{10}$
$\dfrac{6}{18}$	$\dfrac{7}{14}$	$\dfrac{6}{9}$	$\dfrac{9}{15}$
$\dfrac{12}{15}$	$\dfrac{3}{18}$	$\dfrac{10}{12}$	$\dfrac{8}{12}$

Four Aboard

Task   Students read fifteen dominoes as fractions and determine how to arrange them in groups of five having sums of 4.

Setup   A small group of students uses a "Four Aboard" mat and one set of double-six dominoes with the zero-suit and the doubles removed.

Start-up   Explain that a domino may be read as both a *proper fraction* and an *improper fraction*. The 3–2 domino, for example, can be read as the proper fraction ⅔ and as the improper fraction ³⁄₂. Then show how dominoes may be grouped to represent a number sentence whose sum is equal to a whole number, as shown in the example at the right.

$$\frac{3}{2} + \frac{2}{4} = 2$$

Challenge students to use dominoes to model other fraction number sentences having from one to four addends and a whole-number sum, such as 2.

Students position the fifteen dominoes vertically and read them as fractions. Then they arrange the dominoes on their mats in three rows of five so that each row has a sum of 4.

Discussion   Some students will realize that the process of finding addends with a sum of 4 is simplified by finding fractions with like denominators and/or common denominators as well as by finding improper fractions that can be renamed as whole numbers. Other students may notice that for any number sentence it is helpful to first find multiples of 1.

One solution:

Keep up   Have students explore and try to find another whole-number sum, 10 for example, that can be made with the same fifteen dominoes.

One solution: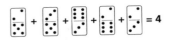

Wrap-up   Key question for discussion or response in journals:
- How did you use like denominators and/or common denominators to find whole-number sums?

Follow-up   Have students add the doubles from 1–1 to 6–6 to the group of fifteen dominoes. Challenge them to arrange all 21 dominoes to form a six-level pyramid in which each level has a different sum from 1 to 6.

$= 4$

$= 4$

$= 4$

```
┌ ─ ─ ─ ─ ┐      ┌ ─ ─ ─ ─ ┐      ┌ ─ ─ ─ ─ ┐
│    |    │      │    |    │      │    |    │
└ ─ ─ ─ ─ ┘      └ ─ ─ ─ ─ ┘      └ ─ ─ ─ ─ ┘
     +                +                +
┌ ─ ─ ─ ─ ┐      ┌ ─ ─ ─ ─ ┐      ┌ ─ ─ ─ ─ ┐
│    |    │      │    |    │      │    |    │
└ ─ ─ ─ ─ ┘      └ ─ ─ ─ ─ ┘      └ ─ ─ ─ ─ ┘
     +                +                +
┌ ─ ─ ─ ─ ┐      ┌ ─ ─ ─ ─ ┐      ┌ ─ ─ ─ ─ ┐
│    |    │      │    |    │      │    |    │
└ ─ ─ ─ ─ ┘      └ ─ ─ ─ ─ ┘      └ ─ ─ ─ ─ ┘
     +                +                +
┌ ─ ─ ─ ─ ┐      ┌ ─ ─ ─ ─ ┐      ┌ ─ ─ ─ ─ ┐
│    |    │      │    |    │      │    |    │
└ ─ ─ ─ ─ ┘      └ ─ ─ ─ ─ ┘      └ ─ ─ ─ ─ ┘
     +                +                +
┌ ─ ─ ─ ─ ┐      ┌ ─ ─ ─ ─ ┐      ┌ ─ ─ ─ ─ ┐
│    |    │      │    |    │      │    |    │
└ ─ ─ ─ ─ ┘      └ ─ ─ ─ ─ ┘      └ ─ ─ ─ ─ ┘
```

Computation

Double-Six Dominoes

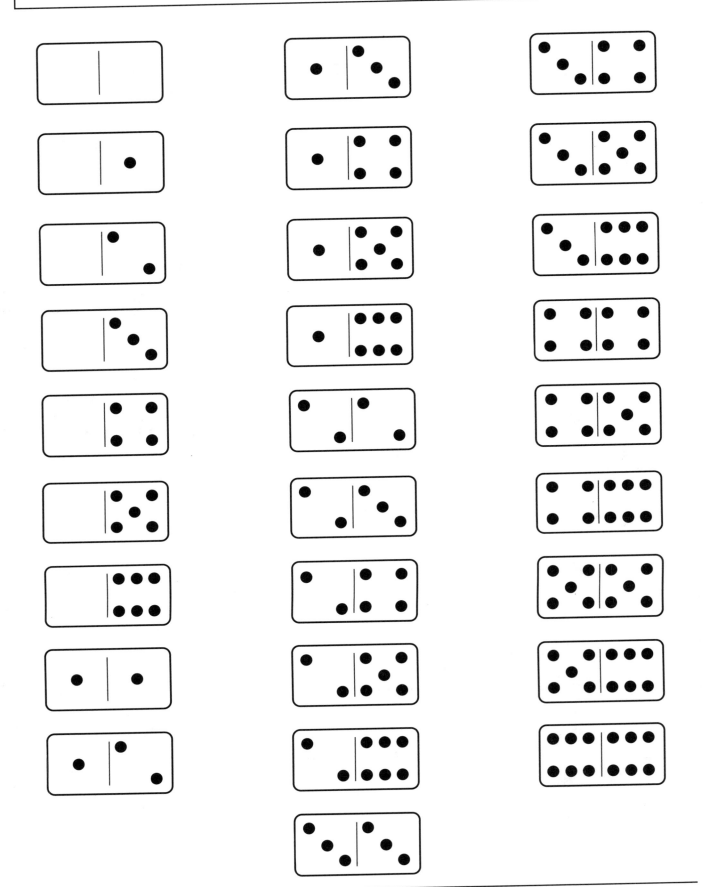

Double-Nine Dominoes

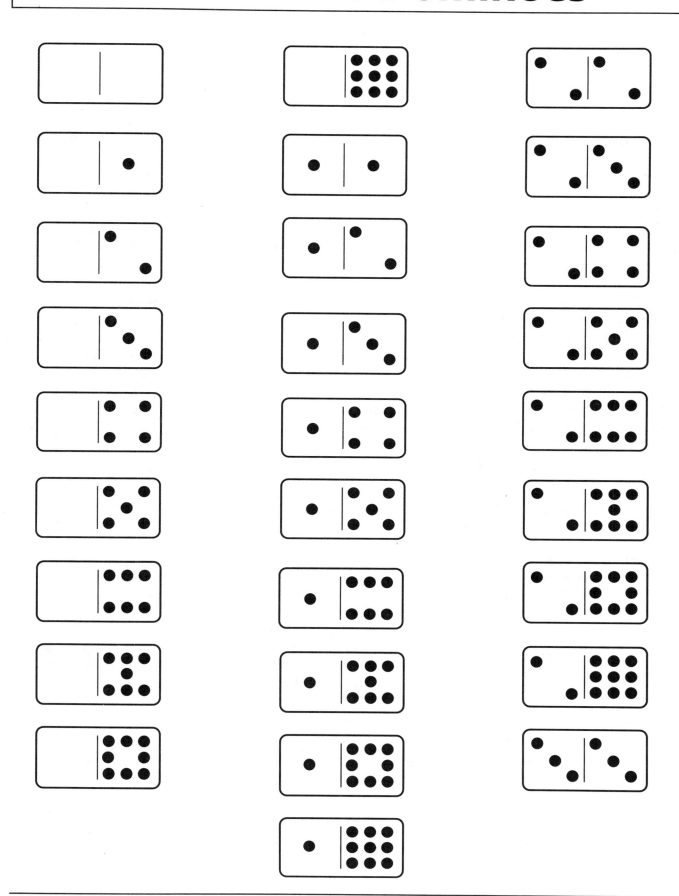

Math Activities with **DOMINOES**
© ETA/Cuisenaire®

Double-Nine Dominoes

(continued)

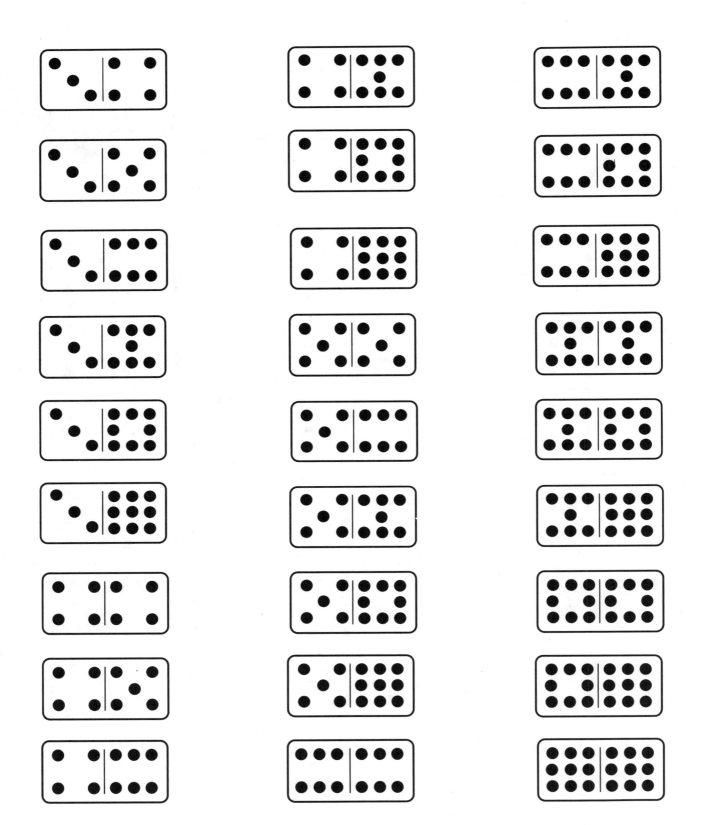